T0335912

Oxidative Stress in Applied Basic Research and Clinical Practice

More information about this series at http://www.springer.com/series/8145

Note from the Editor-in-Chief

All books in this series illustrate point-of-care testing and critically evaluate the potential of antioxidant supplementation in various medical disorders associated with oxidative stress. Future volumes will be updated as warranted by emerging new technology, or from studies reporting clinical trials.

Donald Armstrong
Editor-in-Chief

Martin Rodriguez-Porcel
Alejandro R. Chade • Jordan D. Miller
Editors

Studies on Atherosclerosis

Editors
Martin Rodriguez-Porcel
Cardiovascular Diseases
Mayo Clinic
Rochester, MN, USA

Jordan D. Miller
Department of Surgery
Mayo Clinic
Rochester, MN, USA

Alejandro R. Chade
University of Mississippi
Jackson, MS, USA

ISSN 2197-7224 ISSN 2197-7232 (electronic)
Oxidative Stress in Applied Basic Research and Clinical Practice
ISBN 978-1-4899-7691-8 ISBN 978-1-4899-7693-2 (eBook)
DOI 10.1007/978-1-4899-7693-2

Library of Congress Control Number: 2016944157

Printed on acid-free paper

This Humana Press imprint is published by Springer Nature
The registered company is Springer Science+Business Media LLC New York

Preface

Oxidative stress is a product balance between pro-oxidants and antioxidants. Reactive oxygen species are normally generated by cells and a certain level of oxidative stress plays an important role as second messenger under physiological conditions. However when oxidative stress is increased, whether due to blunted antioxidant defenses or increased generation of reactive oxygen species, it can have a deleterious effect on almost any organ/system in the human. Increased oxidative stress has been shown to promote vasoconstriction, vascular remodeling, inflammation, and fibrosis in different organs. Specifically, as a result of the "oxidative stress modification hypothesis" increased oxidative stress has been directly related to the development of atherosclerosis and vascular disease. Abundant evidence mainly from experimental studies supported a pathophysiological role of increased oxidative stress on the development and progression of vascular disease, which have served as the impetus for many clinical studies to assess the impact of reestablishing oxidant balance in different vascular disease conditions. However, most of them have not provided conclusive or consistent results. Therefore, the relevance of oxidative stress as therapeutic target has been put into question as its pathophysiological role revised.

Many areas of research beyond oxidative stress have had trouble in translating preclinical studies to clinical studies. It is important to make sure that studies, both preclinical and clinical, have the intended hypothesis and are properly designed. Only then, accurate conclusions could be drawn. The goal of this book is to describe a roadmap of the knowledge in the area of oxidative stress and the development of vascular disease and to put in perspective where the scientific field is regarding the translation of preclinical knowledge to clinical trials.

In the first chapters of this book, we will examine the role of oxidative stress under physiological conditions. Subsequent chapters will address the relationship between oxidative stress and the development of vascular disease, and identify and discuss the background that led to clinical trials. The last part of the book will be dedicated to the clinical data that is currently available regarding interventions or attempts to preserve oxidant stress under different conditions. The intention of this last section is to put into perspective the previous clinical studies, what answers they attempted to answer and why they may/may not have tested the intended hypothesis.

It is the hope of the Editors that this book will inspire scientists of many backgrounds to be involved in the field of oxidative stress and vascular disease.

Rochester, MN, USA Martin Rodriguez-Porcel
Jackson, MS, USA Alejandro R. Chade
Rochester, MN, USA Jordan D. Miller

Contents

Contributors

Akshaar Brahmbhatt, M.D. Vascular and Interventional Radiology Translational Laboratory, Mayo Clinic, Rochester, MN, USA

Division of Vascular and Interventional Radiology, Department of Radiology, Mayo Clinic, Rochester, MN, USA

Alastair M. Buchan, F.Med.Sci. Acute Stroke Programme, Radcliffe Department of Medicine, John Radcliffe Hospital, University of Oxford, Oxford, UK

Sophocles Chrissobolis, Ph.D. Department of Pharmacology, Monash University, Clayton, VIC, Australia

Quynh N. Dinh Department of Pharmacology, Monash University, Clayton, VIC, Australia

Grant R. Drummond Department of Pharmacology, Monash University, Clayton, VIC, Australia

Frank M. Faraci, Ph.D. Departments of Internal Medicine and Pharmacology, Francois M. Abboud Cardiovascular Center, Carver College of Medicine, University of Iowa, Iowa City, IA, USA

Iowa City Veterans Affairs Healthcare System, Iowa City, IA, USA

Gina Hadley, M.R.C.P. Acute Stroke Programme, Radcliffe Department of Medicine, John Radcliffe Hospital, University of Oxford, Oxford, UK

Yoshitaka Hirooka, M.D., Ph.D. Department of Advanced Cardiovascular Regulation and Therapeutics, Center for Disruptive Cardiovascular Medicine, Kyushu University, Higashi-ku, Fukuoka, Japan

Danielle Huk Molecular and Cellular Pharmacology Graduate Program, Leonard M. Miller School of Medicine, Miami, FL, USA

Center for Cardiovascular Research at Nationwide Children's Hospital, Research Institute and The Heart Center at Nationwide Children's Hospital, Columbus, OH, USA

Lilach O. Lerman, M.D., Ph.D. Division of Nephrology and Hypertension, Mayo Clinic, Rochester, MN, USA

Joy Lincoln Center for Cardiovascular Research at Nationwide Children's Hospital, Research Institute and The Heart Center at Nationwide Children's Hospital, Columbus, OH, USA

Department of Pediatrics, The Ohio State University, Columbus, OH, USA

Sanjay Misra, M.D., F.S.I.R. F.A.H.A. Vascular and Interventional Radiology Translational Laboratory, Mayo Clinic, Rochester, MN, USA

Division of Vascular and Interventional Radiology, Department of Radiology, Mayo Clinic, Rochester, MN, USA

Ain A. Neuhaus, B.A. Acute Stroke Programme, Radcliffe Department of Medicine, John Radcliffe Hospital, University of Oxford, Oxford, UK

T. Michael De Silva Departments of Internal Medicine and Pharmacology, Francois M. Abboud Cardiovascular Center, Carver College of Medicine, University of Iowa, Iowa City, IA, USA

Iowa City Veterans Affairs Healthcare System, Iowa City, IA, USA

Christopher G. Sobey Department of Pharmacology, Monash University, Clayton, VIC, Australia

Kenji Sunagawa Department of Therapeutic Regulation of Cardiovascular Homeostasis, Center for Disruptive Cardiovascular Medicine, Kyushu University, Fukuoka, Fukuoka, Japan

Xiang-Yang Zhu, M.D., Ph.D. Division of Nephrology and Hypertension, Mayo Clinic, Rochester, MN, USA

About the Authors and Editors

About the Authors

Akshaar Brahmbhatt is currently completing his M.D. at Rutgers University—New Jersey Medical School. His research focuses on the roles of hypoxia, inflammation, oxidative stress, and angiogenesis in the setting of vascular pathologies.

Alastair M. Buchan Professor Buchan is Head of the Medical Sciences Division, Dean of Medicine, Fellow of Corpus Christi College, and Professor of Stroke Medicine at the University of Oxford. He trained at Cambridge, Oxford, and Cornell before moving to Calgary, where he became Director of the Stroke Program. He returned to Oxford in 2005 to establish the thrombolysis service at the John Radcliffe Hospital. His research interests include the role of hamartin in endogenous neuroprotection and pericyte function in models of stroke. Professor Buchan is a Fellow of the Academy of Medical Sciences and an Honorary Member of the American Neurological Association, among other professional accolades.

Sophocles Chrissobolis, Ph.D. is currently an Assistant Professor of Pharmacology in the Department of Pharmaceutical and Biomedical Sciences in the College of Pharmacy at Ohio Northern University (ONU). At ONU, his research is focused on mechanisms that promote hypertension and vascular dysfunction in experimental models of hypertension. His Ph.D. studies at the University of Melbourne focused on mechanisms regulating cerebral vascular function in health and disease. His postdoctoral studies at the University of Iowa, USA, and Monash University, Australia, were supported by fellowships and focused on mechanisms promoting, and protecting against, oxidative stress, vascular dysfunction, and inflammation in experimental hypertension.

Quynh N. Dinh is currently completing her PhD at Monash University. Her research focuses on the roles of aging, inflammation, oxidative stress, and vascular dysfunction in hypertension. She also has an interest in stroke research, particularly on the effects of angiotensin II and aldosterone on stroke outcome. She is currently a member of the High Blood Pressure Research Council of Australia and the American Heart Association.

Grant R. Drummond, Ph.D. is a Senior Research Fellow of the National Health and Medical Research Council of Australia and co-leader (with Professor Chris Sobey) of the Vascular Biology and Immunopharmacology Group in the Cardiovascular Disease Program, Biomedicine Discovery Institute, and Department of Pharmacology, Monash University, Australia. His research interests are on the roles of oxidative and immune mechanisms in the pathophysiology of hypertension, atherosclerosis, and stroke. Grant Drummond received his Ph.D. from the Department of Pharmacology, University of Melbourne, in 1998. He moved to Monash University in 2004 after completing postdoctoral training at Emory University (Atlanta, USA) under Professor David Harrison, and at the Howard Florey Institute (Melbourne, Australia) under Professor Greg Dusting.

Frank M. Faraci received a Ph.D. in Physiology from Kansas State University. He is currently Professor of Internal Medicine and Pharmacology at the University of Iowa. His laboratory focuses on defining molecular mechanisms that underlie large and small vessel disease in brain, with an emphasis on the impact of risk factors for vascular disease—particularly hypertension and aging. He has published over 280 journal articles, served as Associate Editor for *Stroke* and *Arteriosclerosis, Thrombosis, and Vascular Biology (ATVB)* and is currently on the editorial board of the *American Journal of Physiology, Circulation Research*, *Journal of Cerebral Blood Flow and Metabolism, ATVB*, and the *Journal of Neuroscience*. Dr. Faraci's laboratory is funded by the National Institutes of Health, the Department of Veterans Affairs, and the Fondation Leducq.

Gina Hadley Dr. Hadley is an academic clinician with an interest in geriatrics and stroke at the point of submitting her PhD "Mechanisms underlying the endogenous neuroprotection of hamartin in ischaemic stroke" with Professor Alastair Buchan. She has a B.Sc. (Hons.) in Pharmacology with Industrial Experience, spending a year at Queen's University, Canada. She studied Graduate Entry Medicine at the University of Oxford, completing a Wellcome-funded elective project at the Kenya Medical Research Institute (KEMRI). Dr. Hadley is a Tutor in Clinical Medicine at Harris Manchester College, Oxford. She has recently completed a Masters in Evidence Based Medicine and has membership of the Royal College of Physicians.

Yoshitaka Hirooka, M.D., Ph.D. is a professor in the Department of Advanced Cardiovascular Regulation and Therapeutics, at the Kyushu University in Fukuoka, Japan. He is a fellow of the American Heart Association, the European Society of Cardiology, and the Japanese Societies of Cardiology Hypertension. His research interests include neural control of circulation in health and disease state such as hypertension and heart failure, endothelial function, and oxidative stress in cardio-vascular diseases.

Danielle Huk, Ph.D. received her Ph.D. in Molecular and Cellular Pharmacology from the University of Miami, Miller School of Medicine working under Dr. Joy Lincoln. At the University of Miami, and subsequently at Nationwide Children's Hospital in Columbus, OH, Danielle's research centered around defining the molecular mechanisms underlying the onset and progression of heart valve disease. Currently Danielle is working as a postdoctoral researcher at The Ohio State University studying the role of dysregulated Protein Kinase A signaling in tumor formation.

Lilach O. Lerman, M.D., Ph.D. is currently a Professor of Physiology and Medicine and the Director of the Hypertension research at Mayo Clinic. Dr. Lerman obtained her medical degree at the Technion University in Israel and her PhD in Physiology & Biomedical Engineering at Mayo Clinic. Her research is focused on understanding the pathophysiology of hypertension and atherosclerosis, using state-of-the-art imaging modalities. Over the years, she also has had special interest in the role of oxidative stress in kidney disease. More recently, she has integrated her expertise in the study of pathophysiology with the development of novel therapeutics, like cell therapy and others.

Joy Lincoln, Ph.D. is a Principal Investigator at Nationwide Children's Hospital and Associate Professor of Pediatrics at The Ohio State University. Her research program is focused on defining the mechanisms underlying the onset and progression of congenital and acquired heart valve disease. This interest stems from her postdoctoral training under the mentorship of Dr. Katherine Yutzey at Cincinnati Children's Hospital Medical Center. Dr. Lincoln obtained her undergraduate degree in Biomedical Sciences from Durham University in the United Kingdom and graduated with a Ph.D. in Molecular and Developmental Biology from the same institution.

Sanjay Misra, M.D. is a professor in the Division of Vascular/Interventional Radiology, Department of Radiology at Mayo Clinic. Dr. Misra earned his M.D. degree at Hahnemann University School of Medicine and completed a fellowship in cardiovascular and interventional radiology at the Johns Hopkins Hospital. Dr. Misra's clinical practice focuses on peripheral atherosclerotic disease, renal artery stenosis, and dialysis interventions. His research interests include the vascular biology of hemodialysis graft failure, local drug delivery using nanoparticles, regenerative cellular therapies, and novel imaging techniques. Dr. Misra is on the board of the Society of Interventional Radiology and serves on the American Heart Association Council on Cardiovascular Radiology and Interventional Executive Committee, Council on Peripheral Vascular Disease, and Committee on Scientific Sessions Program.

Ain A. Neuhaus obtained his B.A. in Medical Sciences from Somerville College, University of Oxford (1st Class, 2012). He subsequently started a D.Phil. in Medical Sciences with Prof. Alastair Buchan at the Radcliffe Department of Medicine in Oxford, using in vivo and in vitro models of stroke to investigate pericyte contractility and regulation of cerebral blood flow following ischemia and reperfusion, and is now in his final year of D.Phil. studies.

T. Michael De Silva received a PhD in Pharmacology from Monash University in Melbourne, Australia. He completed his postdoctoral training with Dr. Frank M. Faraci at the University of Iowa. Following his postdoctoral training, Dr. De Silva returned to Monash University to establish his own research program. He is the recipient of a National Health and Medical Research Council of Australia Early Career Fellowship. Dr. De Silva's research aims to understand mechanisms that regulate cerebral microvascular function and the impact that microvascular dysfunction has on cognition. Dr. De Silva is an emerging young investigator in the field of cerebrovascular disease and has published 14 journal articles to date.

Chris Sobey, Ph.D. Professor Chris Sobey's research has focused on vascular diseases involving oxidative stress and inflammation, especially in the brain and cerebral circulation. He obtained his Ph.D. at the University of Melbourne for his studies of coronary vascular function following myocardial ischemia. He then completed a postdoctoral Fellowship at the University of Iowa where he gained expertise in the study of cerebral artery function in vivo, and returned to Australia where he is an NHMRC Senior Research Fellow with an established research program in experimental cerebrovascular disease and stroke. His research is now investigating the inflammatory mechanisms occurring in the brain after stroke in order to identify and develop novel approaches to ultimately treat stroke patients.

Kenji Sunagawa, M.D., Ph.D. is a professor in the Center for Disruptive Cardiovascular Medicine and the Department of Cardiovascular Medicine at Kyushu University, Japan. He is a fellow of the American Heart Association, the American College of Cardiology, the European Society of Cardiology, and the Japanese Society of Cardiology. His research interests include Cardiovascular mechanics, Cardiovascular regulation, Heart failure, and Bionic medicine.

Xiang-Yang Zhu, M.D., Ph.D. is a Research Scientist and Assistant Professor at Mayo Clinic, Rochester. His research focuses on the roles of cardiovascular risk factors-related oxidative stress and inflammation on microvascular dysfunction of heart and kidney. He also has an interest in stem cell research, particularly on the effects of mesenchymal stem cells on renovascular disease. He has published several important research articles in journals such as *Circulation*, *the Journal of American Society of Nephrology*, and *Stem Cells*.

About the Editors

Alejandro R. Chade, M.D., F.A.H.A. is a clinically trained Cardiologist who joined the Mayo Clinic following completion of his clinical training for postdoctoral research training in renovascular hypertension and chronic renal disease. Dr. Chade is currently a Tenured Professor in the Department of Physiology and Biophysics at the University of Mississippi Medical Center. His research focuses on the mechanisms of renal injury and therapeutic developments for chronic renovascular disease with a major focus on the renal microcirculation. His research is highly translational and has been published in 70 peer-reviewed papers, book chapters, and is consistently presented in national and international meetings. Dr. Chade is an active member of the American Heart Association, American Physiological Society, American Society of Nephrology, and the European Renal Association.

Jordan D. Miller, Ph.D. is an Associate Professor of Surgery who joined Mayo Clinic in 2009. He received his Ph.D. in Exercise Physiology from the University of Wisconsin-Madison in 2005 and pursued postdoctoral training with Dr. Donald Heistad at the University of Iowa. Dr. Miller's program focuses on mechanisms that contribute to the pathogenesis of age-related cardiovascular diseases, including calcific aortic valve stenosis, atherosclerosis, and stiffening of conduit arteries. His program also aims to bridge the gap from bench to bedside and recently initiated Phase I and II clinical trials aimed at testing the therapeutic efficacy of re-activation of nitric oxide signaling in patients with aortic valve stenosis. He is very active in the American Heart Association and has held several leadership positions in the ATVB Council, is a member of the American Physiological Society, serves on the Board of Directors for the Heart Valve Society, and serves on the Editorial Boards of *Circulation Research*, *American Journal of Physiology-Heart and Circulatory Physiology*, and *the Journal of Heart Valve Disease*.

Martin Rodriguez-Porcel, Ph.D., F.A.H.A is an Associate Professor of Medicine at Mayo Medical School and a consultant in the Department of Cardiovascular Diseases at Mayo Clinic. Dr. Rodriguez-Porcel earned his M.D. degree at the Universidad del Salvador in Buenos Aires, Argentina; Did his internal medicine residency in Alton Ochsner Medical Foundation, New Orleans, LA; and his cardiology fellowship at Mayo Clinic, Rochester, MN. His research interests include the role of oxidative stress in atherosclerosis and the noninvasive imaging of atherosclerosis and regenerative therapies.

Chapter 1
Oxidative Stress in Cardiac Valve Development

Danielle Huk and Joy Lincoln

Abbreviations

aVIC	Activated valve interstitial cell
BMP	Bone morphogenetic protein
ECM	Extracellular matrix
EMT	Endothelial-to-mesenchymal transformation
eNOS	Endothelial nitric oxide signaling
eNOS	Endothelial nitric oxide synthase
ERK	Extracellular-signal-regulated kinase
MAPK	Mitogen-activated protein kinase
MMP	Matrix metalloproteinases
NFATc1	Nuclear factor of activated T-cells (c1)
ROS	Reactive oxygen species
SMA	α-Smooth muscle actin

D. Huk
Molecular and Cellular Pharmacology Graduate Program, Leonard M. Miller School of Medicine, P.O. Box 016189 (R-189), Miami, FL, USA

Center for Cardiovascular Research at Nationwide Children's Hospital Research Institute and The Heart Center at Nationwide Children's Hospital, 575 Children's Crossroads, Research Building III, WB4239, Columbus, OH 43215, USA

J. Lincoln (✉)
Center for Cardiovascular Research at Nationwide Children's Hospital Research Institute and The Heart Center at Nationwide Children's Hospital, 575 Children's Crossroads, Research Building III, WB4239, Columbus, OH 43215, USA

Department of Pediatrics, The Ohio State University, Columbus, OH, USA
e-mail: joy.lincoln@nationwidechildrens.org

M. Rodriguez-Porcel et al. (eds.), *Studies on Atherosclerosis*, Oxidative Stress in Applied Basic Research and Clinical Practice, DOI 10.1007/978-1-4899-7693-2_1

Tgfβ	Transforming Growth Factor β
VEC	Valve endothelial cell
VEGF	Vascular endothelial growth factor
VIC	Valve interstitial cell

Introduction

Oxidative stress is caused by a disturbance in the balance of reactive oxygen species (ROS) generation and attenuated antioxidant defense mechanisms. Minor disturbances likely lead to homeostatic adaptations in response to changes in the immediate environment. While oxidative stress is considered detrimental to the cell, it is becoming recognized that many components of ROS play important roles as secondary messengers in many intracellular signaling pathways [1]. Therefore, homeostatic concentrations could be important for normal physiological processes including valve development. While this area of research is currently understudied, this chapter will focus on the potential role that oxidative stress plays in mediating signaling pathways important for embryonic valvulogenesis and therefore shed light on how alterations in these processes could attribute to valve pathology.

Heart Valve Structure and Function

Heart Valve Function

The average adult's heart beats around 70 times per minute and therefore heart valves must open and close over 100,000 times a day to maintain unidirectional blood flow through the heart. In healthy individuals this is achieved by structure–function relationships between the cellular and extracellular components of the mature valve and the hemodynamic environment. There are two sets of cardiac valves: the mitral and tricuspid atrioventricular valves separate the atria from the ventricles on the left and right sides respectively; and the aortic and pulmonic semilunar valves that separate the ventricles from the great arteries [2]. Although the functional demand of each valve set is similar, their anatomies are different. The atrioventricular valves consist of the mitral and tricuspid valve sets made up of two (mitral) or three (tricuspid) valve leaflets and external supporting chordae tendineae that attach the underside of the valve leaflet to the papillary muscles within the ventricle [3]. The three leaflets of the semilunar valves (aortic, pulmonic) are referred to as cusps and lack external support, although a unique supporting structure is apparent within the aortic roots in the form of a fibrous annulus [4]. It is the coordinated movement of these valvular structures that maintain unidirectional blood flow during the cardiac cycle: In diastole, the papillary muscles are relaxed and high pressure in the atrium causes opening of the mitral and tricuspid valve leaflets to promote blood flow into the respective ventricle. Once ventricular pressure increases during diastole, the chordae "pull back" on the atrioventricular valve leaflets and maintain coaptation to prevent eversion of the valve into the atria. As the ventricle contracts, blood exits through the now open semilunar valves and the ventricle relaxes to begin the cycle again. Therefore, throughout the cardiac cycle the heart valve structures are exposed to constant changes in hemodynamic force as a result of pressure differences between systole to diastole. To withstand this complex mechanical environment, the valve leaflets/cusps develop and maintain an intricate and highly organized connective tissue system that provides all the necessary biomechanical properties for efficient function.

Heart Valve Structure

The mature valve structures are composed of an outer layer of valve endothelial cells (VECs) that surround three stratified layers of specialized extracellular matrix (ECM), interspersed with valve interstitial cells (VICs) (Fig. 1.1) [5–7]. Each ECM layer is organized according to blood flow and collectively they work together to withstand the continual changes in hemodynamic flow during the cardiac cycle [8, 9]. The fibrosa is located on the ventricular side of the atrioventricular (mitral, tricuspid) valve leaflets and atrial side of the semilunar (aortic, pulmonic) cusps, furthest away

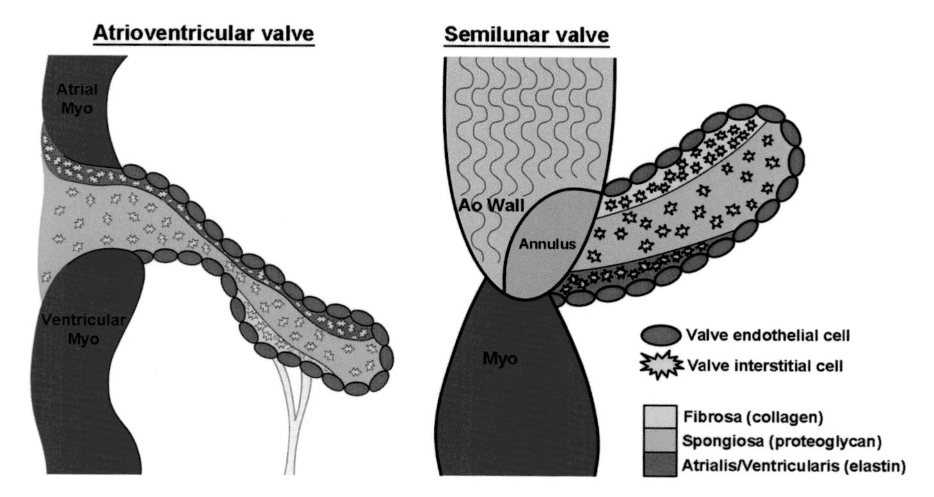

Fig. 1.1 Heart valve structure. Mature heart valves are highly organized structures composed of three stratified layers of extracellular matrix, interspersed with valve interstitial cells (VIC) and surround by a layer of valve endothelial cells (VEC). The fibrosa (*yellow*) is primarily composed of collagens and provides tensile strength; the spongiosa (*blue*) made up of proteoglycans allows compressibility, while the atrialis (*gray*) (atrioventricular valves) or ventricularis (semilunar valves) provides stretch. The ordering of these extracellular matrix layers is relative to blood flow in that the fibrosa is positioned furthest away as indicated by *arrows*. This structure is conserved in both atrioventricular (a) and semilunar (b) valve structures. *Ao Wall* aortic wall, *Myo* myocardium. Image adapted from [7]

from blood flow. This layer is predominantly composed of bundles of collagen fibers aligned along the circumferential direction of the leaflets [9–12]. This arrangement provides tensile strength to the valve leaflets during opening, while transmitting forces to promote coaptation in the closed position [8, 13, 14]. Adjacent to the fibrosa is the spongiosa layer, with a lower abundance of collagens and high prevalence of proteoglycans. This composition provides a more compressible matrix, allowing the valve to geometrically "flex" and absorb high force [9, 15]. Finally, the layer adjacent to blood flow is termed the atrialis (atrioventricular) or ventricularis (semilunar) and largely consists of radially orientated elastin fibers that facilitate tissue movement by extending as the valve leaflet opens and recoiling during closure [2, 16]. Further support is provided by the fibrous annulus structure at the connection between the leaflets/cusps and the myocardium [2, 17], and the external tendinous-chordae in the atrioventricular position. In addition to the ECM, the VIC and VEC populations play essential roles in maintaining connective tissue homeostasis in the functional valve leaflet.

VICs are the most abundant cell type in the valves and distributed throughout all three layers of the ECM [16]. Although originally thought of as a homogenous population of fibroblast-like cells, VICs are now considered highly heterogeneous with quiescent, activated, and progenitor-like phenotypes reported [18]. In healthy adults, VICs are largely quiescent and fibroblast-like and serve to maintain homeostasis of the ECM through a physiological balance of secretion and degradation [7].

It has been shown that VIC function can be influenced by VECs that form an uninterrupted endothelium over the surface of the mature valve cusps. The endothelium is protective and largely serves to sense the surrounding hemodynamic environment and molecularly communicate with underlying VICs to regulate ECM turnover in healthy valves and remodeling in response to injury or mechanical stress [19]. As normal valve function is dependent on the complex arrangement of connective tissue mediated by VICs and VECs, it is not surprising that alterations in cell function and/or ECM distribution leads to inefficiencies and valve disease. Oxidative stress has previously shown to play a significant role in the onset and progression of valve dysfunction and disease [20, 21]; however, to date there are few studies describing the importance in developing valves. By integrating reports describing the role of oxidative stress in mediating signaling pathways with molecular studies of heart valve development, we will discuss the potential role of physiological levels of oxidative stress for valve development, and how aberrations in the embryo could lead to valve disease after birth.

Heart Valve Development and the Potential Role of Oxidative Stress

In the embryo, formation of the heart valves is tightly regulated by a network of active and repressive signaling pathways including those emanating to and from VECs and VICs to establish the tri-laminar structure [7]. Many of these pathways are shared with those affected by high levels of oxidative stress in disease states [22]. It is therefore considered that physiological levels of oxidative stress in the embryo could mediate signal transduction events in these cell populations and modulates components of the ECM during normal valvulogenesis.

Heart Valve Development

Endocardial Cushion Formation

The primitive vertebrate heart tube consists of an outer myocardial cell layer and a continual inner layer of specialized endothelial cells known as the endocardium [23]. The myocardial and endocardial cell layers are separated by specialized ECM, referred to as cardiac jelly [24]. Soon after rightward looping, myocardial cells localized within the atrioventricular canal and outflow tract regions increase deposition of ECM components including proteoglycans and hyaluronan and give rise to cardiac jelly "swellings." Concurrently, a subpopulation of endocardial cells overlying these "swellings" lose contact with neighboring cells as marked by downregulation of E-cadherin, and undergo endothelial-to-mesenchymal transformation (EMT) [24, 25]. As a result, newly transformed mesenchyme cells invade the underlying cardiac jelly and rapidly proliferate to populate the endocardial cushions [24–26] (Fig. 1.2, left). This essential role of VECs during early stages of valvulogenesis is initiated in the chick around

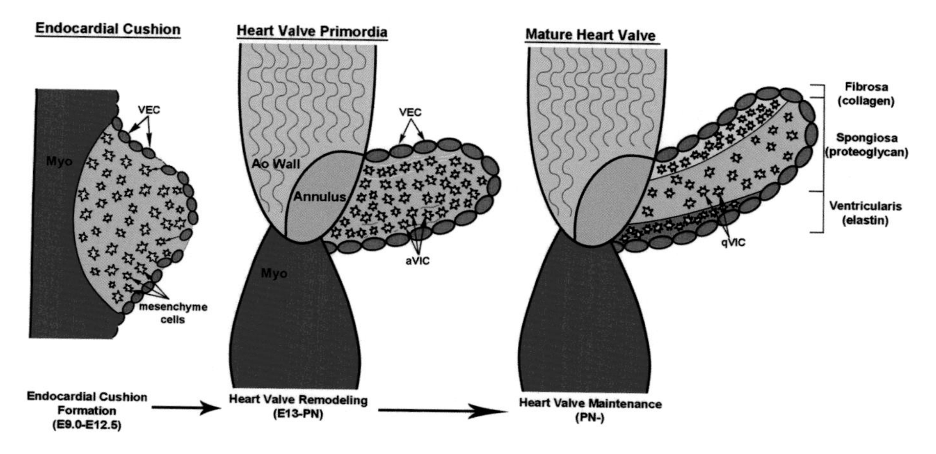

Fig. 1.2 Overview of heart valve development. (*Left*) Heart valve development is initiated in the embryo with endocardial cushion formation following transformation of valve endothelial cells (VECs) into mesenchyme cells in the atrioventricular canal and outflow tract regions (outflow tract shown). (*Middle*) The endocardial cushion swellings elongate and mesenchyme cells mature into valve interstitial cells (VICs) that promote remodeling of the valve primorida. Activated VICs (aVICs) are highly synthetic and secrete extracellular matrix proteins that give rise to the fibrosa (*yellow*), spongiosa (*blue*), and ventricularis (*gray*) layers of the mature valve (*Right*). Soon after birth, VICs become quiescent (qVICs) and the mature valve structure is maintained throughout life. *Ao Wall* aortic wall, *Myo* myocardium. Time line indicated is relative to mouse gestation. Image adapted from [7]

embryonic day (E) 3, in the mouse at ~E9.0 and between E31-E35 in human development [27]. In the atrioventricular canal, four endocardial cushions form: the superior, inferior, and left and right lateral cushions; while in the outflow tract, two cushions form in proximal and distal locations [28]. These cushion structures serve as physical barriers to prevent backflow of blood through the primitive heart tube [29], in addition to hosting a population of mesenchyme cells that serve as precursors to the mature valve structures [30, 31]. Our lab and others have described this population of mesenchyme cells as pluripotent based on their potential to differentiate into multiple cell lineages in response to molecular cues in vitro [32]. Initially these cells express typical mesenchyme cell markers including *twist1*, *msx2,* and *tbx20* [26] as well as high levels of SMA and are therefore referred to as primitive VICs during these early stages.

In the atrioventricular canal, the majority of primitive VICs are endothelially derived [30]. However cells from the epicardium have also been shown to populate the cushions and parietal leaflets of the atrioventricular valves [33–36]. In the avian system, quail-chick transplantation models identified epicardial-derived cells as a source of VIC precursors [34]; however the chicken pro-epicardium does not show significant valve cell investment, consistent with other lineage mapping studies in the mature avian valves [31]. In mice, genetic approaches to fate map epicardial cell lineages using *Wt1*- and *Tbx18-Cre* have reported *Cre*-positive cells within mature valve leaflets of the mouse, primarily within the parietal leaflets [35–37]. Therefore, the contribution of epicardial cells in the formation of mature valve structures is likely species dependent. In the outflow tract, epicardial-derived cells are not detected, but neural crest and secondary heart field-derived cells are observed [38–40]. These studies support the conclusion that the atrioventricular and semilunar valves are derived from several cell lineages, however the specific function of each of these is not clear.

Heart Valve Remodeling

Once the VIC precursor pool has been expanded in the developing cushions by the multiple cell derivatives, EMT halts; however the VECs remain as an intact endothelium over the developing valve surface and regain cell contact. At this time, the inferior and superior cushions in the atrioventricular canal fuse and elongate to give rise to the septal leaflets and the lateral cushions begin to form the mural leaflets [30, 31] (Fig. 1.2, middle). Similar morphological changes occur in the outflow tract cushions in parallel with septation of the aorta and pulmonary artery by the conotruncal ridges [24]. This occurs around E14.5 in the mouse and E6 in the chick. Around this time, the primitive VICs lose expression of mesenchyme markers, but maintain high levels of SMA expression. This state of "VIC activation" is associated with degradation of the hyaluronan components of the endocardial cushion and secretion of ECM components that will later form the three layers of the mature valve. During this stage, VIC proliferation remains active, but is significantly less than during EMT [32, 41]. Remodeling of the valve leaflets continues through birth and soon after, the tri-laminar layers of the valve become evident. Coinciding with this, VICs lose expression of SMA and become quiescent and fibroblast-like, and remain this way throughout life in the absence of disease [7].

Signaling Pathways of Heart Valve Development

Valvulogenesis is tightly regulated by the interaction of multiple signaling pathways, many of which are affected by ROS in pathological processes. Therefore the potential role of physiological oxidative stress in regulating stages of valve development is discussed.

Signaling Pathways required for Endocardial Cushion Formation

Transformation of VECs into mesenchyme cells is essential for endocardial cushion formation during early stages of valvulogenesis. Endothelial-to-mesenchymal transformation (EMT) is common in the development of several embryonic tissues and active in the process of pathologic tumor metastasis. The regulatory mechanisms of this process in health and disease are shared and involve components of ROS signaling and other pathways mediated by ROS including Transforming Growth Factor (Tgf) β, Wnt, and Vascular endothelial growth factor (VEGF). The requirement of these pathways in heart valve development and their potential connection with ROS are discussed (Fig. 1.3).

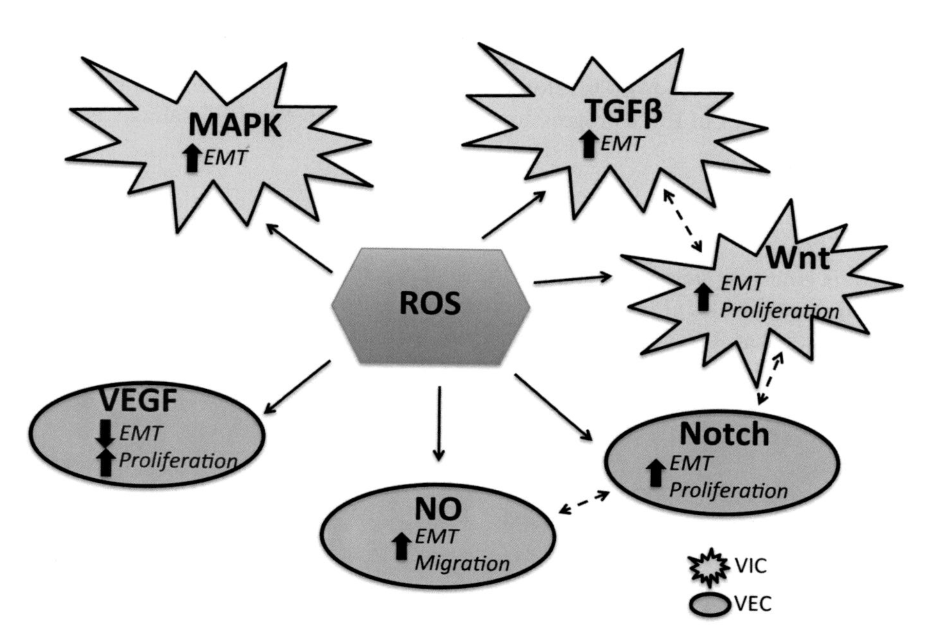

Fig. 1.3 Signaling pathways potentially affected by reactive oxygen species that mediate function of VECs (*red*) and VICs (*yellow*) during normal endocardial cushion formation. *VIC* valve interstitial cell, *VEC* valve endothelial cell

Endothelial Nitric Oxide Synthase Signaling

As the name suggests, endothelial nitric oxide synthase (eNOS) plays an important role in the valve endothelial cells during development. In embryos null for *eNOS*, EMT is initiated but mesenchyme cell transformation and migration are decreased in cushion explant assays [42]. Therefore suggesting that although endothelial, this secreted enzyme acts on adjacent newly transformed mesenchyme cells. While this concept is new to valve development, studies have suggested a similar molecular communication between these neighboring cells and paracrine actions of eNOS in adult VICs during pathological processes [19, 43–45]. During embryonic EMT, eNOS is activated and dependent on Notch signaling via PI3-kinase/Akt [46]. Similar to the $eNOS^{-/-}$ model [42], embryos lacking Notch components initiate EMT, but fail to form populated endocardial cushions [47], supporting studies by Bosse et al., suggesting molecular interaction between these two pathways [44]. $eNOS^{-/-}$ mice have varied longevity seemingly dependent upon strain, with one study reporting 38.3 % lethality soon after birth [48, 49]. However endocardial cushion defects were not reported in this study and premature lethality was reportedly due to septal defects [48]. Interestingly, two independent studies have reported the abnormal formation of two, instead of three aortic valve cusps in $eNOS^{-/-}$ mice at birth [44, 50], suggesting that this congenital condition arises due to developmental abnormalities as a result of eNOS deletion. However, it is not yet known if defects in EMT and endocardial cushion formation underlie bicuspid aortic valve (BAV) phenotypes. The penetrance of BAV in $eNOS^{-/-}$ mice is somewhat low and between 23 and 36 % [44, 50], however this is significantly increased (91 %) when bred into the $Notch1^{+/-}$ background [44]. Therefore suggesting that eNOS plays an important role during later stages of EMT progression and interacts with Notch signaling.

Tgfβ Signaling

There is emerging data from noncardiovascular studies to suggest that nitric oxide signaling mediates Tgfβ-dependent cellular processes [51–53]. In addition, others have shown that ROS is upstream and activates latent Tgfβ and induces *Tgfβ* gene expression [54–64]. In healthy embryos, Tgfβ signaling derived from the myocardium is required in endothelial cells for EMT and endocardial cushion formation. In the avian system, Tgfβ2 and Tgfβ3 are the most potent ligands required for EMT initiation in VECs [65, 66]. In particular, Tgfβ2 mediates VEC cell–cell activation and separation by promoting loss of cell–cell contacts, while Tgfβ3-activating TgfβRII promotes invasion of transformed mesenchyme cells into the cardiac jelly [66]. While in the mouse, loss of *Tgfβ2* does not affect EMT initiation, but developing cushions are variably hypoplastic [67, 68]. Similar phenotypes are also observed in other mouse models with targeted downregulation of Tgfβ signaling including endocardial-deletion of *TgfβRI* (*Alk5*) [69] and global deletion of the long form of latent Tgfβ binding protein 1, in which EMT initiation is impaired [70].

Downstream, Tgfβ signals through noncanonical and canonical pathways to regulate EMT in the developing valves. Increased ERK activity is a known inducer of valvular EMT [71–73], and ROS has been shown to activate the MAPK pathway in other systems [74]. In addition, Tgfβ can function through canonical Smad2/3 in the cushions to promote expression of Snai1 (Snail) and Snai2 (Slug); two-zinc finger transcription factors essential for EMT and highly expressed in VECs undergoing EMT, and newly transformed mesenchyme cells [75–77]. While canonical Smad signaling has been well studied during this process during valvulogenesis, little attention has been paid to the potential role of ROS as an intracellular mediator of Tgfβ-induced EMT. Interestingly, Tgfβ treatment of MDCK cells has been found to increase intracellular levels of ROS, Smad2 phosphorylation, and SMA expression, while decreasing fibronectin and E-cadherin; molecular hallmarks of EMT [78]. Furthermore, this process is dependent on NOX activity. In lung and renal epithelial cells, inhibition of cellular ROS signaling by antioxidants blocks Tgfβ-induced EMT [79, 80]. In addition, hypoxic conditions increase *Tgfβ1* and promote EMT in human malignant mesothelioma cells, leading to an aggressive phenotype and poor prognosis [81]. In addition to the convergence of ROS and Tgfβ signaling for EMT, several studies have implicated that ROS regulates EMT in human bronchial epithelial and breast adenocarcinoma cell lines by stabilizing Snai1, which in turn will promote EMT by repressing *E-cadherin* and causing cell–cell contacts to be lost [82–84]. In another study, ROS enhanced EMT hypermethylation of the *E-cadherin* promoter [85]. Together, these studies highlight the potential role for ROS in Tgfβ-mediated cellular processes involved in EMT and endocardial cushion formation.

Wnt Signaling

Canonical Wnt/β-catenin signaling has been implicated in the regulation of EMT and cell proliferation during endocardial cushion formation [24, 86–88]. In several developmental systems, Wnt/β-catenin signaling can be tightly controlled by ROS as increased Wnt function increases ROS through Nox1 [89]. In endocardial cushions, Wnt-mediated EMT is largely mediated through crosstalk with Tgfβ [24, 86–88]; however, interactions with Notch signaling have also been described as Notch1-mediated EMT requires Wnt4 [90]. Studies have shown that the convergence of Notch and Wnt/β-catenin pathways can in part be regulated by Nox-derived ROS [91, 92], however this may be cell-type dependent and conservation of these mechanisms in endothelial cells of the developing valves remains elusive.

Vascular Endothelial Growth Factor (VEGF) Signaling

While Tgfβ, Notch, and Wnt are all important inducers of EMT, exogenous treatment of vascular endothelial growth factor A (VEGF-A) prevents EMT in collagen gel endocardial cushion explant assays in vitro [93, 94]. VEGF is broadly expressed throughout endocardial cells of the developing heart and becomes restricted to the atrioventricular canal region soon after the onset of EMT [93]. Under normal physiological conditions, VEGF regulates proliferation of endothelial cells to maintain the overlying endothelium as select cells undergo EMT, suggesting a maintenance role. VEGF signals through receptors on the surface of VECs to mediate intracellular signaling pathways including MAPK and NFATc1 and its expression is robustly increased under hypoxic conditions [94]. In addition to its role in VECs, VEGF is also critical in vascular endothelial cells for the formation of new vessels. In this system, there is evidence that ROS could also play a role as VEGF-induced angiogenesis and neovascularization is impaired in the absence of Nox2 [95, 96]. The VEGF receptor type 1 (VEGF-R1) is highly expressed in VECs of the primitive cushion, however it is downregulated after EMT when VEGF-R2 increases to regulate valve elongation [97]. The upstream regulation of VEGF-R2 during this process remains elusive, but studies have shown that excess ROS is sufficient to induce aberrant phosphorylation, even in the absence of the VEGF ligand [98]. These studies show that VEGF levels must be tightly regulated during heart valve development and findings from other systems suggest a role for ROS in mediating this regulatory hierarchy.

Signaling Pathways Implicated in Heart Valve Remodeling

Once endocardial cushion formation is complete, overlying VECs stop undergoing EMT, regain cell–cell contacts, and form an uninterrupted endothelium (Fig. 1.4) [24]. Around this time, the individual endocardial cushions fuse and elongate into mitral and tricuspid valve primordia in the atrioventricular canal region, and primitive aortic and pulmonic structures in the outflow tract [30]. Concurrently, cells

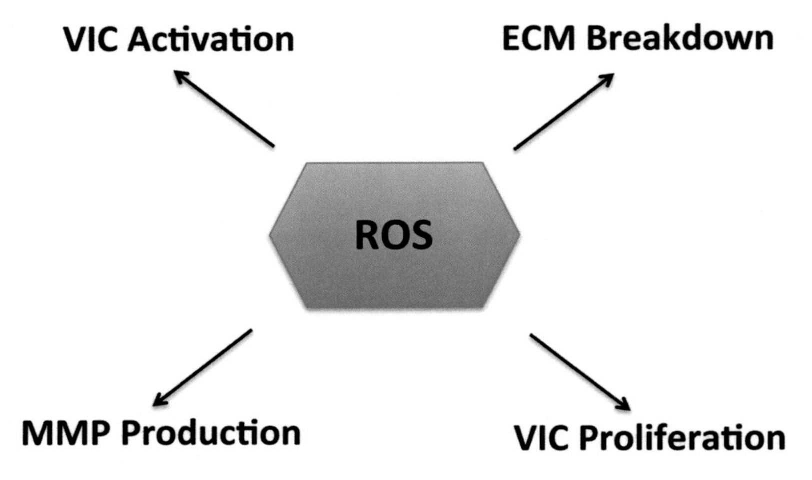

Fig. 1.4 Biological processes potentially affected by reactive oxygen species during normal embryonic heart valve remodeling

within the developing valve structures continue to proliferate and the connective tissue undergoes extensive remodeling. Compared to EMT, much less is known about the regulation of valve remodeling. Work by our group and others has shown that during mid-stages of valvulogenesis VICs lose expression of mesenchyme genes including *Twist1*, *Tbx20*, *Msx1*, and *Msx2*, but maintain SMA expression indicative of a transition towards an activated VIC (aVIC) phenotype [32, 99–101]. During this time, aVICs express high levels of matrix metalloproteinases (MMPs) that break down the primitive ECM within the remodeling cushions. Concurrently, these cells become highly synthetic and begin to secrete specialized ECM including collagens and proteoglycans within the developing stratified structure [16, 102] It is not until postnatal stages that the elastin fibers are laid down in the atrialis/ventricularis layer and the valve becomes tri-stratified [41]. A significant reduction in cell proliferation is another feature of late embryonic valve development, while normally little, if any, cell proliferation occurs after birth [41, 103].

VIC Activation

A key feature of valve remodeling is the maturation of mesenchyme precursor cells towards an activated, myofibroblast-like VIC phenotype. Once the tri-laminar structure of the valve is established soon after birth, the reverse occurs and aVICs transition towards a quiescent VIC (qVIC) phenotype similar to fibroblasts, and cells remain this way throughout life in the absence of disease [16, 101]. The mechanisms that regulate VIC activation and quiescence in the developing valve are not well defined, but insights are provided from studies examining myofibroblast activation in other systems. The most potent activator of this process is Tgfβ and in postnatal VIC, Tgfβ treatment promotes SMA expression [104, 105]. Although the signaling mediators of Tgfβ-dependent VIC activation have not yet been identified, convincing data in other systems suggest that ROS play a critical role as biological second messengers in myofibroblast differentiation [51, 106]. Nox4 is activated by Tgfβ signaling in vascular smooth muscle cells and fibroblasts and leads to increased SMA expression, indicative of a myofibroblast phenotype [107, 108]. In many systems this is associated with pathological fibrosis and not physiological activation of fibroblast-like cells during development, however *Nox4* mRNA has been reported in heart valve structures [107–110]. Therefore, in addition to EMT, ROS-mediated Tgfβ signaling could play a role in VIC plasticity in post-EMT remodeling valves.

Extracellular Matrix Remodeling

Once VICs are activated during remodeling stages, they express mRNAs associated with matrix degradation including MMPs and become highly synthetic [12, 16]. These events facilitate the breakdown of primitive ECM within the endocardial cushion and the secretion of collagens, proteoglycans, and elastins that later form the fibrosa, spongiosa, and ventricularis layers respectively within the mature valve.

At present, little is known of the signaling pathways required for ECM remodeling in the embryo or diseased valves, but oxidative stress has been shown to have multifactorial effects on ECM integrity, effecting ECM proteins both directly and indirectly. Many of the examples reported in the literature describe the role of excess levels of oxidative stress associated with pathological ECM remodeling [109, 111, 112]; however it is considered that physiological levels could be active in developing valves and important for ECM homeostasis.

In vitro studies have suggested that oxidants can directly lead to shedding and fragmentation of ECM components that are found in early valve structures, including heparin sulfate and hyaluronan [113, 114]. The shedding of such proteins has been linked to induction of profibrotic signaling in other systems that promotes expression of ECM proteins such as collagens and proteoglycans by active myofibroblasts [115, 116]. In addition to direct effects, reduced response to oxidative stress can lead to overproduction of collagens [117]. Moreover, decreases in homeostatic levels of nitric oxide have been shown to promote myofibroblast activation and increase production of ECM within the valve in disease [21, 118]. As there is increasing evidence to suggest that developmental pathways are reactivated in valve pathology, it is plausible to consider similar mechanisms are important during valvulogenesis, albeit more controlled.

In addition to ECM secretion in the developing valve, activated myofibroblasts also mediate degradation and turnover through expression of MMPs, whose activities have been shown to be induced by ROS [119]. During valve remodeling, MMPs promote breakdown of the primitive endocardial cushion matrix but also facilitate cell migration during earlier EMT stages [24]. The presence of ROS is thought to promote aberrant MMP production leading to ECM disruption and tissue degradation [120]. While these studies describing the role of oxidative stress in ECM homeostasis are not valve specific, they do provide appreciable insights into potential mechanisms mediated by physiological oxidative stress during valvulogenesis.

Conclusions and Perspectives

Heart valve development is a very complex process integrating many biological processes regulated by intricate networks of signaling pathways. While the role of oxidative stress and reactive oxygen species has not been implicated in valve development there are shared parallels with other systems in which it plays a significant role in regulating signaling pathways. Although, we generally associate oxidative stress as a pathological mediator at high levels, it is considered that the developing valves may be exposed to physiological levels that control signaling pathways important for endocardial cushion formation and valve remodeling. Heart valve phenotypes in embryos and young mice carrying mutations in genes important in anti-oxidative and pro-oxidant responses have not been described [121], but close examination would be highly informative.

Heart valve disease was once considered a degenerative process that increased with age. However, there are emerging reports to suggest that developmental defects underlie valve malformations and dysfunction observed after birth or later in life [102]. With this in mind, it has become increasingly important to understand the key regulators of normal valve development in order to understand disease mechanisms and improve patient outcomes. This review has highlighted the previously unappreciated role(s) that oxidative stress could play in regulating valvulogenesis. Therefore, changes in the balance between ROS generation and antioxidant defense mechanisms could damage cell components or have detrimental effects on the signaling pathways critical for normal valve formation in the embryo, and thereby serve as a potential therapeutic target to reduce the penetrance of congenital cardiac defects in high-risk patient populations.

References

1. Burton GJ, Jauniaux E. Oxidative stress. Best Pract Res Clin Obstet Gynaecol. 2011;25(3):287–99.
2. Hinton RB, Yutzey KE. Heart valve structure and function in development and disease. Annu Rev Physiol. 2011;73:29–46.
3. Anderson RH, Ho SY, Becker AE. Anatomy of the human atrioventricular junctions revisited. Anat Rec. 2000;260(1):81–91.
4. Anderson RH. Clinical anatomy of the aortic root. Heart. 2000;84(6):670–3.
5. Garcia-Martinez V, Sanchez-Quintana D, Hurle JM. Histochemical and ultrastructural changes in the extracellular matrix of the developing chick semilunar heart valves. Acta Anat. 1991;142(1):87–96.
6. Gross L, Kugel MA. Topographic anatomy and histology of the valves in the human heart. Am J Pathol. 1931;7(5):445–74.
7. Tao G, Kotick JD, Lincoln J. Heart valve development, maintenance, and disease: the role of endothelial cells. Curr Top Dev Biol. 2012;100:203–32.
8. Balachandran K, Sucosky P, Yoganathan AP. Hemodynamics and mechanobiology of aortic valve inflammation and calcification. Int J Inflamm. 2011;2011:263870.
9. Lincoln J, Lange AW, Yutzey KE. Hearts and bones: shared regulatory mechanisms in heart valve, cartilage, tendon, and bone development. Dev Biol. 2006;294(2):292–302.
10. Icardo JM, Colvee E. Atrioventricular valves of the mouse: III. Collagenous skeleton and myotendinous junction. Anat Rec. 1995;243(3):367–75.
11. Kunzelman KS et al. Differential collagen distribution in the mitral valve and its influence on biomechanical behaviour. J Heart Valve Dis. 1993;2(2):236–44.
12. Rabkin-Aikawa E, Mayer Jr JE, Schoen FJ. Heart valve regeneration. Adv Biochem Eng Biotechnol. 2005;94:141–79.
13. Aldous IG et al. Differences in collagen cross-linking between the four valves of the bovine heart: a possible role in adaptation to mechanical fatigue. Am J Physiol Heart Circ Physiol. 2009;296(6):H1898–906.
14. Grande-Allen KJ, Liao J. The heterogeneous biomechanics and mechanobiology of the mitral valve: implications for tissue engineering. Curr Cardiol Rep. 2011;13(2):113–20.
15. Sacks MS, David Merryman W, Schmidt DE. On the biomechanics of heart valve function. J Biomech. 2009;42(12):1804–24.
16. Schoen FJ. Evolving concepts of cardiac valve dynamics: the continuum of development, functional structure, pathobiology, and tissue engineering. Circulation. 2008;118(18): 1864–80.

17. Misfeld M, Sievers HH. Heart valve macro- and microstructure. Phil Trans R Soc Lond B Biol Sci. 2007;362(1484):1421–36.
18. Liu AC, Gotlieb AI. Characterization of cell motility in single heart valve interstitial cells in vitro. Histol Histopathol. 2007;22(8):873–82.
19. Butcher JT, Nerem RM. Valvular endothelial cells regulate the phenotype of interstitial cells in co-culture: effects of steady shear stress. Tissue Eng. 2006;12(4):905–15.
20. Roos CM et al. Transcriptional and phenotypic changes in aorta and aortic valve with aging and MnSOD deficiency in mice. Am J Physiol Heart Circ Physiol. 2013;305(10): H1428–39.
21. Miller JD et al. Dysregulation of antioxidant mechanisms contributes to increased oxidative stress in calcific aortic valvular stenosis in humans. J Am Coll Cardiol. 2008;52(10): 843–50.
22. Heistad DD et al. Novel aspects of oxidative stress in cardiovascular diseases. Circ J. 2009;73(2):201–7.
23. Perez-Pomares JM, Gonzalez-Rosa JM, Munoz-Chapuli R. Building the vertebrate heart - an evolutionary approach to cardiac development. Int J Dev Biol. 2009;53(8-10):1427–43.
24. Person AD, Klewer SE, Runyan RB. Cell biology of cardiac cushion development. Int Rev Cytol. 2005;243:287–335.
25. Eisenberg LM, Markwald RR. Molecular regulation of atrioventricular valvuloseptal morphogenesis. Circ Res. 1995;77(1):1–6.
26. Combs MD, Yutzey KE. Heart valve development: regulatory networks in development and disease. Circ Res. 2009;105(5):408–21.
27. Lopez-Sanchez C, Garcia-Martinez V. Molecular determinants of cardiac specification. Cardiovasc Res. 2011;91(2):185–95.
28. Markwald RR et al. Developmental basis of adult cardiovascular diseases: valvular heart diseases. Ann N Y Acad Sci. 2010;1188:177–83.
29. Schroeder JA et al. Form and function of developing heart valves: coordination by extracellular matrix and growth factor signaling. J Mol Med. 2003;81(7):392–403.
30. Lincoln J, Alfieri CM, Yutzey KE. Development of heart valve leaflets and supporting apparatus in chicken and mouse embryos. Dev Dyn. 2004;230(2):239–50.
31. de Lange FJ et al. Lineage and morphogenetic analysis of the cardiac valves. Circ Res. 2004;95(6):645–54.
32. Lincoln J, Alfieri CM, Yutzey KE. BMP and FGF regulatory pathways control cell lineage diversification of heart valve precursor cells. Dev Biol. 2006;292(2):292–302.
33. Wessels A et al. Epicardially derived fibroblasts preferentially contribute to the parietal leaflets of the atrioventricular valves in the murine heart. Dev Biol. 2012;366(2):111–24.
34. Gittenberger-de Groot AC et al. Epicardium-derived cells contribute a novel population to the myocardial wall and the atrioventricular cushions. Circ Res. 1998;82(10):1043–52.
35. Cai CL et al. A myocardial lineage derives from Tbx18 epicardial cells. Nature. 2008;454(7200):104–8.
36. Zhou B et al. Epicardial progenitors contribute to the cardiomyocyte lineage in the developing heart. Nature. 2008;454(7200):109–13.
37. Lockhart MM et al. The epicardium and the development of the atrioventricular junction in the murine heart. J Dev Biol. 2014;2(1):1–17.
38. Nakamura T, Colbert MC, Robbins J. Neural crest cells retain multipotential characteristics in the developing valves and label the cardiac conduction system. Circ Res. 2006;98(12):1547–54.
39. Jiang X et al. Fate of the mammalian cardiac neural crest. Development. 2000; 127(8):1607–16.
40. Mjaatvedt CH et al. Normal distribution of melanocytes in the mouse heart. Anat Rec A Discov Mol Cell Evol Biol. 2005;285(2):748–57.
41. Hinton Jr RB et al. Extracellular matrix remodeling and organization in developing and diseased aortic valves. Circ Res. 2006;98(11):1431–8.

42. Liu Y et al. Nitric oxide synthase-3 promotes embryonic development of atrioventricular valves. PLoS One. 2013;8(10), e77611.
43. El Accaoui RN et al. Aortic valve sclerosis in mice deficient in endothelial nitric oxide synthase. Am J Physiol Heart Circ Physiol. 2014;306(9):H1302–13.
44. Bosse K et al. Endothelial nitric oxide signaling regulates Notch1 in aortic valve disease. J Mol Cell Cardiol. 2013;60:27–35.
45. Gould ST et al. The role of valvular endothelial cell paracrine signaling and matrix elasticity on valvular interstitial cell activation. Biomaterials. 2014;35(11):3596–606.
46. Chang AC et al. Notch initiates the endothelial-to-mesenchymal transition in the atrioventricular canal through autocrine activation of soluble guanylyl cyclase. Dev Cell. 2011;21(2):288–300.
47. Timmerman LA et al. Notch promotes epithelial-mesenchymal transition during cardiac development and oncogenic transformation. Genes Dev. 2004;18(1):99–115.
48. Feng Q et al. Development of heart failure and congenital septal defects in mice lacking endothelial nitric oxide synthase. Circulation. 2002;106(7):873–9.
49. Shesely EG et al. Elevated blood pressures in mice lacking endothelial nitric oxide synthase. Proc Natl Acad Sci U S A. 1996;93(23):13176–81.
50. Lee TC et al. Abnormal aortic valve development in mice lacking endothelial nitric oxide synthase. Circulation. 2000;101(20):2345–8.
51. Sampson N, Berger P, Zenzmaier C. Redox signaling as a therapeutic target to inhibit myofibroblast activation in degenerative fibrotic disease. Biomed Res Int. 2014;2014:131737.
52. Grubisha MJ, DeFranco DB. Local endocrine, paracrine and redox signaling networks impact estrogen and androgen crosstalk in the prostate cancer microenvironment. Steroids. 2013; 78(6):538–41.
53. Samarakoon R, Overstreet JM, Higgins PJ. TGF-beta signaling in tissue fibrosis: redox controls, target genes and therapeutic opportunities. Cell Signal. 2013;25(1):264–8.
54. Barcellos-Hoff MH. Latency and activation in the control of TGF-beta. J Mammary Gland Biol Neoplasia. 1996;1(4):353–63.
55. Amarnath S et al. Endogenous TGF-beta activation by reactive oxygen species is key to Foxp3 induction in TCR-stimulated and HIV-1-infected human CD4+CD25− T cells. Retrovirology. 2007;4:57.
56. Jobling MF et al. Isoform-specific activation of latent transforming growth factor beta (LTGF-beta) by reactive oxygen species. Radiat Res. 2006;166(6):839–48.
57. Pociask DA, Sime PJ, Brody AR. Asbestos-derived reactive oxygen species activate TGF-beta1. Lab Invest. 2004;84(8):1013–23.
58. Sullivan DE et al. The latent form of TGFbeta(1) is induced by TNFalpha through an ERK specific pathway and is activated by asbestos-derived reactive oxygen species in vitro and in vivo. J Immunotoxicol. 2008;5(2):145–9.
59. Vodovotz Y et al. Regulation of transforming growth factor beta1 by nitric oxide. Cancer Res. 1999;59(9):2142–9.
60. Wang H, Kochevar IE. Involvement of UVB-induced reactive oxygen species in TGF-beta biosynthesis and activation in keratinocytes. Free Radic Biol Med. 2005;38(7):890–7.
61. Leonarduzzi G et al. The lipid peroxidation end product 4-hydroxy-2,3-nonenal up-regulates transforming growth factor beta1 expression in the macrophage lineage: a link between oxidative injury and fibrosclerosis. FASEB J. 1997;11(11):851–7.
62. Bellocq A et al. Reactive oxygen and nitrogen intermediates increase transforming growth factor-beta1 release from human epithelial alveolar cells through two different mechanisms. Am J Respir Cell Mol Biol. 1999;21(1):128–36.
63. Saito K et al. Iron chelation and a free radical scavenger suppress angiotensin II-induced upregulation of TGF-beta1 in the heart. Am J Physiol Heart Circ Physiol. 2005; 288(4):H1836–43.
64. Shvedova AA et al. Increased accumulation of neutrophils and decreased fibrosis in the lung of NADPH oxidase-deficient C57BL/6 mice exposed to carbon nanotubes. Toxicol Appl Pharmacol. 2008;231(2):235–40.

65. Nakajima Y et al. Mechanisms involved in valvuloseptal endocardial cushion formation in early cardiogenesis: roles of transforming growth factor (TGF)-beta and bone morphogenetic protein (BMP). Anat Rec. 2000;258(2):119–27.
66. Mercado-Pimentel ME, Runyan RB. Multiple transforming growth factor-beta isoforms and receptors function during epithelial-mesenchymal cell transformation in the embryonic heart. Cells Tissues Organs. 2007;185(1-3):146–56.
67. Sanford LP et al. TGFbeta2 knockout mice have multiple developmental defects that are non-overlapping with other TGFbeta knockout phenotypes. Development. 1997;124(13): 2659–70.
68. Bartram U et al. Double-outlet right ventricle and overriding tricuspid valve reflect disturbances of looping, myocardialization, endocardial cushion differentiation, and apoptosis in TGF-beta(2)-knockout mice. Circulation. 2001;103(22):2745–52.
69. Sridurongrit S et al. Signaling via the Tgf-beta type I receptor Alk5 in heart development. Dev Biol. 2008;322(1):208–18.
70. Todorovic V et al. Long form of latent TGF-beta binding protein 1 (Ltbp1L) regulates cardiac valve development. Dev Dyn. 2011;240(1):176–87.
71. Lee MW et al. The involvement of reactive oxygen species (ROS) and p38 mitogen-activated protein (MAP) kinase in TRAIL/Apo2L-induced apoptosis. FEBS Lett. 2002;512(1-3): 313–8.
72. Benhar M et al. Enhanced ROS production in oncogenically transformed cells potentiates c-Jun N-terminal kinase and p38 mitogen-activated protein kinase activation and sensitization to genotoxic stress. Mol Cell Biol. 2001;21(20):6913–26.
73. Krenz M et al. Role of ERK1/2 signaling in congenital valve malformations in Noonan syndrome. Proc Natl Acad Sci U S A. 2008;105(48):18930–5.
74. Son Y et al. Mitogen-activated protein kinases and reactive oxygen species: how can ROS activate MAPK pathways? J Signal Transduct. 2011;2011:792639.
75. Cho HJ et al. Snail is required for transforming growth factor-beta-induced epithelial-mesenchymal transition by activating PI3 kinase/Akt signal pathway. Biochem Biophys Res Commun. 2007;353(2):337–43.
76. Romano LA, Runyan RB. Slug is an essential target of TGFbeta2 signaling in the developing chicken heart. Dev Biol. 2000;223(1):91–102.
77. Tao G et al. Mmp15 is a direct target of Snai1 during endothelial to mesenchymal transformation and endocardial cushion development. Dev Biol. 2011;359(2):209–21.
78. Zhang A, Dong Z, Yang T. Prostaglandin D2 inhibits TGF-beta1-induced epithelial-to-mesenchymal transition in MDCK cells. Am J Physiol Renal Physiol. 2006;291(6): F1332–42.
79. Gorowiec MR et al. Free radical generation induces epithelial-to-mesenchymal transition in lung epithelium via a TGF-beta1-dependent mechanism. Free Radic Biol Med. 2012; 52(6):1024–32.
80. Rhyu DY et al. Role of reactive oxygen species in TGF-beta1-induced mitogen-activated protein kinase activation and epithelial-mesenchymal transition in renal tubular epithelial cells. J Am Soc Nephrol. 2005;16(3):667–75.
81. Kim MC, Cui FJ, Kim Y. Hydrogen peroxide promotes epithelial to mesenchymal transition and stemness in human malignant mesothelioma cells. Asian Pac J Cancer Prev. 2013;14(6):3625–30.
82. Chen F et al. Loss of Ikkbeta promotes migration and proliferation of mouse embryo fibroblast cells. J Biol Chem. 2006;281(48):37142–9.
83. Dong R et al. Stabilization of snail by HuR in the process of hydrogen peroxide induced cell migration. Biochem Biophys Res Commun. 2007;356(1):318–21.
84. Cano A et al. The transcription factor snail controls epithelial-mesenchymal transitions by repressing E-cadherin expression. Nat Cell Biol. 2000;2(2):76–83.
85. Lim SO et al. Epigenetic changes induced by reactive oxygen species in hepatocellular carcinoma: methylation of the E-cadherin promoter. Gastroenterology. 2008;135(6):2128–40. e1–8.

86. Hurlstone AF et al. The Wnt/beta-catenin pathway regulates cardiac valve formation. Nature. 2003;425(6958):633–7.
87. Liebner S et al. Beta-catenin is required for endothelial-mesenchymal transformation during heart cushion development in the mouse. J Cell Biol. 2004;166(3):359–67.
88. Alfieri CM et al. Wnt signaling in heart valve development and osteogenic gene induction. Dev Biol. 2010;338(2):127–35.
89. Kajla S et al. A crucial role for Nox 1 in redox-dependent regulation of Wnt-beta-catenin signaling. FASEB J. 2012;26(5):2049–59.
90. Wang Y et al. Endocardial to myocardial notch-wnt-bmp axis regulates early heart valve development. PLoS One. 2013;8(4), e60244.
91. Coant N et al. NADPH oxidase 1 modulates WNT and NOTCH1 signaling to control the fate of proliferative progenitor cells in the colon. Mol Cell Biol. 2010;30(11):2636–50.
92. Funato Y, Miki H. Redox regulation of Wnt signalling via nucleoredoxin. Free Radic Res. 2010;44(4):379–88.
93. Dor Y et al. A novel role for VEGF in endocardial cushion formation and its potential contribution to congenital heart defects. Development. 2001;128(9):1531–8.
94. Armstrong EJ, Bischoff J. Heart valve development: endothelial cell signaling and differentiation. Circ Res. 2004;95(5):459–70.
95. Ushio-Fukai M et al. Novel role of gp91(phox)-containing NAD(P)H oxidase in vascular endothelial growth factor-induced signaling and angiogenesis. Circ Res. 2002;91(12):1160–7.
96. Tojo T et al. Role of gp91phox (Nox2)-containing NAD(P)H oxidase in angiogenesis in response to hindlimb ischemia. Circulation. 2005;111(18):2347–55.
97. Stankunas K et al. VEGF signaling has distinct spatiotemporal roles during heart valve development. Dev Biol. 2010;347(2):325–36.
98. Domigan CK, Ziyad S, Iruela-Arispe ML. Canonical and noncanonical vascular endothelial growth factor pathways: new developments in biology and signal transduction. Arterioscler Thromb Vasc Biol. 2015;35(1):30–9.
99. Chakraborty S et al. Twist1 promotes heart valve cell proliferation and extracellular matrix gene expression during development in vivo and is expressed in human diseased aortic valves. Dev Biol. 2010;347(1):167–79.
100. Hurle JM et al. Elastic extracellular matrix of the embryonic chick heart: an immunohistological study using laser confocal microscopy. Dev Dyn. 1994;200(4):321–32.
101. Rabkin E et al. Activated interstitial myofibroblasts express catabolic enzymes and mediate matrix remodeling in myxomatous heart valves. Circulation. 2001;104(21):2525–32.
102. Lincoln J, Yutzey KE. Molecular and developmental mechanisms of congenital heart valve disease. Birth Defects Res A Clin Mol Teratol. 2011;91(6):526–34.
103. Aikawa E et al. Human semilunar cardiac valve remodeling by activated cells from fetus to adult: implications for postnatal adaptation, pathology, and tissue engineering. Circulation. 2006;113(10):1344–52.
104. Pho M et al. Cofilin is a marker of myofibroblast differentiation in cells from porcine aortic cardiac valves. Am J Physiol Heart Circ Physiol. 2008;294(4):H1767–78.
105. Hinz B. Formation and function of the myofibroblast during tissue repair. J Invest Dermatol. 2007;127(3):526–37.
106. Forman HJ et al. The chemistry of cell signaling by reactive oxygen and nitrogen species and 4-hydroxynonenal. Arch Biochem Biophys. 2008;477(2):183–95.
107. Cucoranu I et al. NAD(P)H oxidase 4 mediates transforming growth factor-beta1-induced differentiation of cardiac fibroblasts into myofibroblasts. Circ Res. 2005;97(9):900–7.
108. Hecker L et al. NADPH oxidase-4 mediates myofibroblast activation and fibrogenic responses to lung injury. Nat Med. 2009;15(9):1077–81.
109. Hagler MA et al. TGF-beta signalling and reactive oxygen species drive fibrosis and matrix remodelling in myxomatous mitral valves. Cardiovasc Res. 2013;99(1):175–84.
110. Liu RM, Gaston Pravia KA. Oxidative stress and glutathione in TGF-beta-mediated fibrogenesis. Free Radic Biol Med. 2010;48(1):1–15.

111. Hulin A et al. Emerging pathogenic mechanisms in human myxomatous mitral valve: lessons from past and novel data. Cardiovasc Pathol. 2013;22(4):245–50.
112. Miller JD et al. Lowering plasma cholesterol levels halts progression of aortic valve disease in mice. Circulation. 2009;119(20):2693–701.
113. Gao F et al. Extracellular superoxide dismutase inhibits inflammation by preventing oxidative fragmentation of hyaluronan. J Biol Chem. 2008;283(10):6058–66.
114. Kliment CR et al. Oxidative stress alters syndecan-1 distribution in lungs with pulmonary fibrosis. J Biol Chem. 2009;284(6):3537–45.
115. Gauldie J et al. Transfer of the active form of transforming growth factor-beta 1 gene to newborn rat lung induces changes consistent with bronchopulmonary dysplasia. Am J Pathol. 2003;163(6):2575–84.
116. Kliment CR, Oury TD. Oxidative stress, extracellular matrix targets, and idiopathic pulmonary fibrosis. Free Radic Biol Med. 2010;49(5):707–17.
117. Phillippi JA et al. Altered oxidative stress responses and increased type I collagen expression in bicuspid aortic valve patients. Ann Thorac Surg. 2010;90(6):1893–8.
118. Rajamannan NM et al. Atorvastatin inhibits calcification and enhances nitric oxide synthase production in the hypercholesterolaemic aortic valve. Heart. 2005;91(6):806–10.
119. Fu X et al. Hypochlorous acid generated by myeloperoxidase modifies adjacent tryptophan and glycine residues in the catalytic domain of matrix metalloproteinase-7 (matrilysin): an oxidative mechanism for restraining proteolytic activity during inflammation. J Biol Chem. 2003;278(31):28403–9.
120. Grote K et al. Mechanical stretch enhances mRNA expression and proenzyme release of matrix metalloproteinase-2 (MMP-2) via NAD(P)H oxidase-derived reactive oxygen species. Circ Res. 2003;92(11):e80–6.
121. Pouyet L, Carrier A. Mutant mouse models of oxidative stress. Transgenic Res. 2010; 19(2):155–64.

Chapter 2
ROS in Atherosclerotic Renovascular Disease

Xiang-Yang Zhu and Lilach O. Lerman

Introduction

Atherosclerotic renovascular disease (ARVD) is a major cause of secondary hypertension, especially in the elderly population. ARVD typically involves the proximal third of the renal artery including the perirenal aorta and ostium. The prevalence of ARVD is about 6.3–6.8 % in patients undergoing angiography or as determined in a population-based study utilizing a noninvasive screening technique [1, 2], and is increased in elderly patients with additional comorbid conditions such as diabetes, hypertension, or coronary artery disease [3]. ARVD is a progressive disease in regards to both the lesion severity and kidney function, and might confer a poor prognosis to affected patients.

The relationship between unilateral renal artery stenosis and arterial hypertension has been demonstrated in the original experiments of Goldblatt et al. [4], which led to the discovery of renin-angiotensin-aldosterone system (RAAS). Furthermore, because of the reduced perfusion beyond the stenosis, the tissue of the stenotic kidney is exposed to chronic hypoxia, which leads to ischemic kidney disease. In the contralateral kidney however, renal damage progressively develops by the hypertension caused by activation of the RAAS and volume overload.

We have previously shown in our animal model an early increase in systemic plasma renin activity (PRA, 4-5 weeks after induction of renal artery stenosis), as seen in the early phases of renovascular hypertension [5, 6], returns to baseline by 8 weeks. Few data track PRA over time in ARVD patients, but it seems to also show an early increase and later normalize [7]. These data indicate that hypertension in

X.-Y. Zhu, M.D., Ph.D. • L.O. Lerman, M.D., Ph.D. (✉)
Division of Nephrology and Hypertension, Mayo Clinic, 200 First Street SW, Rochester, MN 55905, USA
e-mail: Lerman.Lilach@Mayo.Edu

M. Rodriguez-Porcel et al. (eds.), *Studies on Atherosclerosis*, Oxidative Stress in Applied Basic Research and Clinical Practice, DOI 10.1007/978-1-4899-7693-2_2

the late phase of ARVD may be maintained by other mechanisms. We have shown [8] that increased oxidative stress and upregulation of inflammatory factors in ARVD are associated with marked impairments of renal hemodynamics and function and that increased abundance of reactive oxygen species (ROS) initially leads to renal microvascular endothelial dysfunction, which may precede and subsequently be aggravated by the development of obstructive lesions in the main renal artery. In a murine model of renovascular hypertension [9], up regulation of both pro- and anti-oxidant genes were observed as early as 3 days after renal artery stenosis before the renal tissue demonstrates any histologic abnormalities. Oxidative stress is defined as a tissue injury induced by increase in ROS such as oxygen radicals, which can be generated at different sites along the nephron, like the glomeruli and segments two and three of the proximal tubule. ARVD is associated with activation of oxidative pathways, reduction in nitric oxide (NO) synthesis and stimulation of the RAAS, in both human subjects and experimental models (Fig. 2.1). The goal of this chapter is to review the potential role of ROS in ARVD.

Mechanisms of ROS Generation in the Stenotic Kidney

RAAS Activation

More than one century ago, Tigerstedt and Bergman introduced the concept that "renin" in the kidney may have a systemic pressor effect [10]. In the 1930s, Goldblatt et al. [4] demonstrated that uni- or bilateral clip of the renal artery cause blood pressure rise and attributed the pressor effect to "internal secretion" of the kidney that triggers vasoconstriction. Renin was subsequently successfully isolated by Braun-Menendez and Irvine Page [11], but renin itself has little vasopressor activity. Skeggs and colleagues [12] characterized the hormonal cascade downstream to renin, including angiotensin-I (Ang I), angiotensin converting enzyme (ACE), and angiotensin-II (Ang II). Kidney juxtaglomerular (JG) cells secrete renin, which converts angiotensinogen (the main source of which is the liver) to Ang I. Ang I is cleaved by ACE originating from the lung and kidney to Ang II, which elevates blood pressure by several mechanisms. Ang II activates sympathetic nerve activity and induces arteriolar vasoconstriction, but also increases sodium and water retention by tubular reabsorption, which is accelerated by aldosterone secreted by the adrenal gland. Furthermore, Ang II stimulates anti-diuretic hormone secretion in the pituitary gland, leading to increased water reabsorption in the collecting duct [13]. Those systemic effects of Ang II increase water and sodium retention and the effective circulating volume, resulting in increased perfusion pressure at the JG apparatus, leading to decreased renin secretion by negative feedback. Therefore, the decrease in renal perfusion pressure induced by the vascular occlusion in ARVD stimulates JG cells to secrete renin and thereby activates the cascade of RAAS resulting in renovascular hypertension.

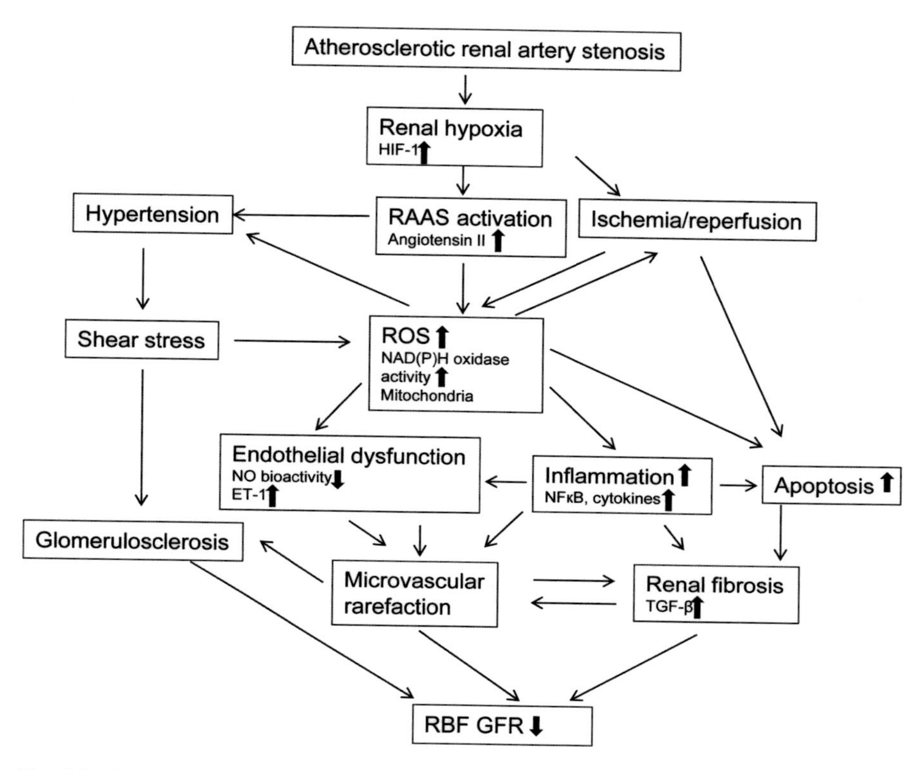

Fig. 2.1 Overview of the role of reactive oxygen species (ROS) in the mechanisms of atherosclerotic renovascular disease (ARVD). In ARVD, renal ischemia induces tissue hypoxia and hypoxia inducible factor (HIF)-1 upregulation. Hypoxia activates renin-angiotensin-aldosterone system (RAAS). Angiotensin II mainly stimulate nicotinamide adenine dinucleotide phosphate (NAD(P)H)-oxidase to generate ROS, which decrease nitric oxide (NO) bioactivity and induce endothelial dysfunction. Angiotensin II also induces inflammatory factors and cytokines such as nuclear factor kappa B (NFκB) and transforming growth factor (TGF)-β, and cell apoptosis, subsequently lead to microvascular rarefaction and renal fibrosis. Furthermore, angiotensin II, ROS, and inflammation impair tubular function, lead water sodium retention elevate blood pressure. Hypertension-induced shear stress can also induce ROS impair endothelial function and cause glomerulosclerosis. These entire cascades together decrease renal blood flow (RBF) and glomerular filtration rate (GFR)

In addition to systemic RAAS, all the elements of RAAS have been discovered and can be produced in the kidney [14]. Renin is secreted from JG cells, the collecting duct, and collecting duct. Intrarenal angiotensinogen is produced mainly in the proximal tubule and ACE in the proximal tubule, the collecting duct, and endothelial cells. Furthermore, intrarenal formation of Ang II is independent of the circulating RAAS. The concentration of intrarenal Ang II is much higher than systemic level. In an early phase of renal artery stenosis, the JG cellular renin is increased in the stenotic kidney. After the acute stage, JG renin is suppressed by negative feedback, upregulated renin is produced in the distal nephron and in the collecting duct in the contralateral kidney to support continued intrarenal Ang II formation leading to maintenance of hypertension. Furthermore, Ang II activates proximal tubular secretion of ACE and ACE binding activity [15].

Ang II-induced activation of nicotinamide adenine dinucleotide phosphate (NAD(P)H) oxidase leads to increased production of ROS. Ang II simulated mitochondrial NAD(P)H oxidase isoform 4 and resulted in the abrupt production of mitochondrial superoxide and hydrogen peroxide (H_2O_2). Ang II also induced depolarization of the mitochondrial membrane potential, and cytosolic secretion of cytochrome C and apoptosis-inducing factor [16] (Fig. 2.1).

Elevated ROS produce an oxidized form of angiotensinogen with greater potency to bind renin, facilitates angiotensin release, and increase blood pressure [17]. Since ROS are extracellular signaling molecules, they may be significant in mediating the actions of vasoconstrictors, such as Ang II, thromboxane A2, endothelin-1, adenosine, and norepinephrine. Thus, the activation RAAS starts the entire ROS-dependent cascades in ARVD.

Inflammation

The activation of the RAAS leads to production of Ang II, which can induce oxidative stress by upregulating NAD(P)H oxidase, impairs endothelial function, and increases vascular permeability, subsequently increase macrophage infiltration [18], possibly by upregulation of monocyte chemoattractant protein-1 (MCP-1), a chemokine that increases monocyte infiltration into inflamed tissues and an important inflammatory mediator. In mice ischemia/reperfusion (I/R) model, there was a large influx of inflamed monocytes into the kidney. These monocytes produced tumor necrosis factor (TNF)-alpha, interleukin (IL)-6, IL-1α, and IL-12 [19]. Activated macrophages and fibroblasts in the ARVD kidney may again directly induce NAD(P)H oxidase activity, stimulating transforming growth factor (TGF)-β1 production and triggering fibroblast proliferation and differentiation into collagen-secreting myofibroblasts [20]. Neutralization of MCP-1 reduces macrophage infiltration and progressive kidney damage in rat tubulointerstitial nephritis [21]. Acute infusion with Ang II significantly increases leukocytes adhesion in the rat mesenteric arteries and increased expression of vascular cell adhesion molecule (VCAM)-1 in rat aorta via nuclear factor kappa-B (NF-κB) transcriptional activation [22], which can be inhibited by administration of losartan, an Ang II type I Receptors (AT1R) antagonist. Studies show that Ang II elicits proinflammatory responses in the kidney by regulating the expression of cytokines and chemokines [23]. Ang II induces NF-κB activation and the expression of IL-6 in human vascular smooth muscle cells [24]. Ang II also play a significant role in the initiation and progression of atherogenesis [25], an inflammation mediated process, where Ang II provides a positive feedback loop in vascular inflammation via recruitment of inflammatory cells, which then induce production of more Ang II, therefore perpetuating vascular inflammation. Moreover, inflammatory factors like TNF-α can also induce ROS that serve as second messengers for intracellular signaling, in mesangial cells, TNF-α induces apoptosis through superoxide anion, but not H_2O_2 [26].

The proinflammatory effects of the RAAS are partly mediated by aldosterone. Aldosterone promotes insulin resistance and vascular remodeling and influences the development of atherosclerosis. Chronic infusion of aldosterone induces oxidative stress in rat aorta, and MR antagonist spironolactone reduces ROS generation. Human and rat adrenal cortical cells stimulated with Ang II produce aldosterone via AT1R-upregulation of cytochrome P450 oxidase B2 and increased level of hydrogen peroxide, whereas pretreatment with losartan and antioxidants abrogates Ang II effects [27].

These studies suggest the interaction between ROS and inflammation in the kidney injury in ARVD.

Comorbidities

Coexistence of hypercholesterolemia aggravates the effects of renal artery stenosis on both vascular and kidney injury. For example, unilateral renal artery narrowing in ApoE/mice results in chronic vascular inflammation and accelerated atherosclerosis compared to sham-surgery [28]. The atheroma is cellular in composition and stains for the presence of macrophages and MCP-1 in both the distal abdominal aorta and carotid artery [28].

Furthermore, our previous study [8] demonstrated that the combination of hypercholesterolemia and renal artery stenosis amplifies activation of mechanisms that can promote renal vascular, glomerular, and tubulointerstitial injury compared with renal artery stenosis alone. Hypercholesterolemia and renal artery stenosis were associated with a marked increase in tubular and glomerular expression of profibrotic TGF-β, tissue inhibitor of matrix metalloproteinase (TIMP)-1, and (plasminogen activator inhibitor (PAI)-1, accompanied by increased expression of NF-κB, but attenuated the expression of MMP-2 and ubiquitin, and decreased apoptosis compared with ARVD, suggesting a shift in the tissue remodeling process that favors renal fibrosis and matrix accumulation.

Diabetes is a common risk factor that often coexists with ARVD. Increased oxidative stress, formation of advanced glycoxidation end products (AGEs), chronic inflammation, and activated cellular response are the major molecular mechanisms of atherogenesis in diabetic patients. Furthermore, elevated free fatty acids, high glucose levels, or AGEs induce ROS in vascular cells, leading to ongoing AGE formation and to generation of proinflammatory cytokines [29]. Moreover, elevated cytokines in obesity and diabetes may also induce oxidative stress thus a vicious cycle may be initiated and accelerated. Increased ROS may upregulate NF-κB expression through protein kinase C and the mitogen-activated protein kinase pathway, and subsequently induce the expression of numerous cytokines which act on vascular cells promoting the deleterious effects [30]. Studies have demonstrated that the NAD (P)H oxidase and the AGE/RAGE/NF-κB axis accelerated atherosclerosis [31]. The blockade of ROS or AGE formation at different sites may interrupt the vicious cycle, e.g. ACE inhibitors, AT1 receptor blockers, 3-hydroxy-3- methyl-glutaryl-CoA reductase inhibitors (statins), and thiazolidindiones have shown intracellular antioxidant activity in addition to their primary pharmacological actions.

Elements of Ischemia-Reperfusion Injury

Kidney ischemia-reperfusion (I/R) injury (IRI) is often observed in acute kidney injury, which impairs renal function through different cascades of ROS and inflammation compared with chronic ARVD (Fig. 2.2). However, it may also occur in ARVD due to cholesterol emboli or rupture of an atherosclerotic lesion causing an

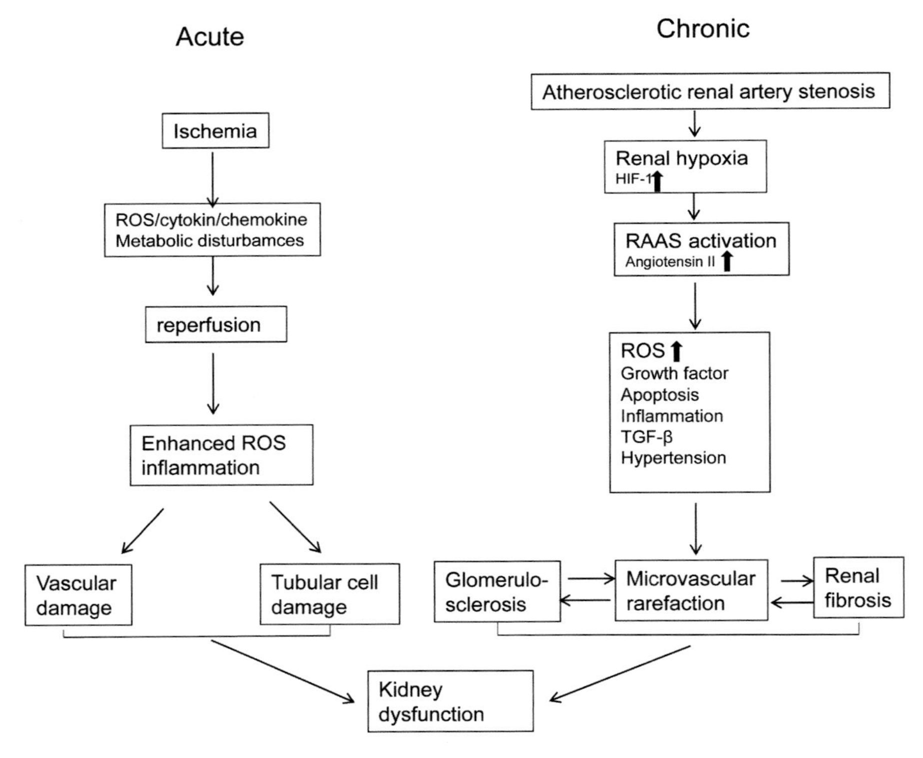

Fig. 2.2 Acute vs. chronic modulation of renal function in atherosclerotic renovascular disease (ARVD). In acute renal ischemia, interruption of kidney perfusion results in a rapid drop in oxygen and nutrient supply, which leads to hypoxic damage and ROS formation. Reperfusion induces a massive and local production of ROS and inflammatory factors, which are responsible for vascular and tubular cell damage. In chronic ARVD, the mechanisms are series cascades of activation renin-angiotensin-aldosterone system (RAAS), including ROS, inflammation apoptosis and growth factor degrading, eventually lead to glomerulosclerosis, microvascular rarefaction, and renal fibrosis

in site thrombosis. Initially, a sustained interruption of kidney perfusion results in a rapid drop in oxygen and nutrient supply, which leads to hypoxic damage. Hypoxia causes a rapid depletion of the energy supply, intracellular accumulation of lactate, and acidification of cell cytosol. In addition, the lack of energy delivery induces disorganization of the cytoskeleton, disruption of intercellular tight junctions, loss of cell polarity, and dysfunction of membrane ion transporters. Consequently, epithelial and endothelial cells detach from their basal membrane and obstruct tubular and vascular lumens. The loss of tubular and vascular select-permeability eventuates accumulation of fluids in the interstitium, which further delays kidney reperfusion and prolongs the ischemic insult. Ischemia also induced ROS which cause cellular injury. However, abrupt reperfusion also induces a massive and local production of ROS, which are responsible for detrimental oxidation of proteins, lipids, membranes, and nucleic acids of both epithelial and endothelial cells, leads to pronounced tissue damage [32]. In addition to all these metabolic consequences, inflammatory reaction characterized by the expression and activation of endothelial adhesion molecules, integrins, and selectins also participate the processing of IRI [33]. These inflammatory molecules in turn activate the innate immune responses via the Toll-like receptors, and recruit inflammatory cells [34]. The deleterious impact of I/R-associated inflammation and infiltration of monocytes involves chemokine receptors, such as chemokine receptor-2, chemokine receptor-7, and CXC chemokine receptor-4, as well as the local production of ROS, TNF-α, and IL-1β [35].

Recent studies demonstrate that microRNAs, which are short endogenous non-coding RNAs, are important regulators of target messenger RNAs involved in IRI. Dicer cleaves pre-microRNA into short microRNA. Targeted deletion of Dicer from the proximal tubular epithelium protects from I/R-induced renal injury (preserved renal function, blunted tissue damage and tubular apoptosis, and survival benefit) and is associated with changes in the expression of distinct microRNAs (e.g., miR-132, -362, and -379) [36]. Furthermore, microvesicles secreted by endothelial progenitor cells and enriched with microRNA (including miR-126 and miR-296) were shown to ameliorate IRI in the murine kidney [37].

Taken together, ROS dominated in I/R-induced kidney injury.

Effects of ROS on the Stenotic Kidney

Vascular Function (Renal Perfusion and Endothelial Function)

A significant renal artery stenosis leads to persistent reduction of renal parenchymal perfusion and eventually may lead to loss of kidney function. In mild renal artery stenosis, renal perfusion may be sustained if the post-stenotic pressure remains within the range of autoregulation [38]. However, with increasing severity of stenosis, renal adaptation reaches its limit and a fall in renal blood flow ensues. In chronic ARVD, the reduction of renal blood flow triggers multiple mechanisms of tissue and vascular injury that lead to progressive renal fibrosis. Activation of RAAS induces

Ang II-dependent release of ROS, which trigger cascades of inflammatory processes and perivascular/interstitial fibrosis, eventually cause irreversible kidney damage [39]. Sympathetic overactivity may further activate RAAS, enhancing oxidative stress [40]. Endothelial function is disturbed through an imbalance between vasoconstrictors (mainly endothelin) and vasodilators (such as NO and prostacyclin), and plays pivotal role in the pathogenesis of ARVD (Fig. 2.1). Moreover, ROS directly cause renal vasoconstriction, which in turn decreases renal blood flow and function.

Superoxide anion rapidly scavenges nitric oxide NO and could therefore blunt NO activity in the renal vasculature, induces endothelial dysfunction, which is the initial step in the pathogenesis of atherosclerosis and contributes to development of hypertension. Renovascular hypertension may be secondary to excess Ang II and increased oxidative stress [5], but in turn the mechanical stretch in hypertension may provoke an increase in oxidative stress [41]. After renal angioplasty, forearm blood flow response to acetylcholine was enhanced in renovascular hypertensive patients, suggesting improvement in endothelial function. The increase in maximal acetylcholine-induced vasodilation was associated with decrease in urinary excretion of 8- hydroxy-2′-deoxyguanosine (OHdG) and in serum concentration of malondialdehyde-low density lipoprotein (MDA-LDL) [42]. Co-infusion of the antioxidant vitamin C augmented acetylcholine-induced vasodilation before angioplasty but not after angioplasty [42]. Figure 2.3 show comparable findings in the stenotic kidney of pig ARVD, in which increased superoxide anion production (Dihydroethidium (DHE) staining) was associated with decreased renal blood flow and its response to acetylcholine, which was improved by anti-oxidant treatment. These findings suggest that endothelial function is impaired in both experimental models and a clinical setting, and that this impairment is at least in part caused by increased oxidative stress.

Tubular Integrity

The key feature of ischemic nephropathy is renal atrophy, which is precipitated mainly by tubular atrophy. In a rat renal artery partial clip model [43], progressive reduction in renal mass was observed during the evolution of renal injury. In the acute phase (2–8 days), both apoptosis and necrosis-induced cell death, but increased tubular epithelial cell labeling and mitoses indicated epithelial repair. In the chronic phase (10–28 days), when the mass of the ischemic kidney underwent significant reduction, only apoptosis contributed to cell death, and the level of tubular epithelial cell labeling and mitosis returned to near normal. Intratubular macrophages were observed to remove the apoptotic bodies. The area of the tubular epithelium was reduced in atrophic tubules, possibly due to apoptotic cell deletion, as well as cell shrinkage. A form of kidney hibernation with readily reversible tubular atrophy has been described in rat with renal artery stenosis induced by a 0.2-mm clip around the left renal artery [44]. Following 7 weeks of clipping and 2 concomitant weeks of

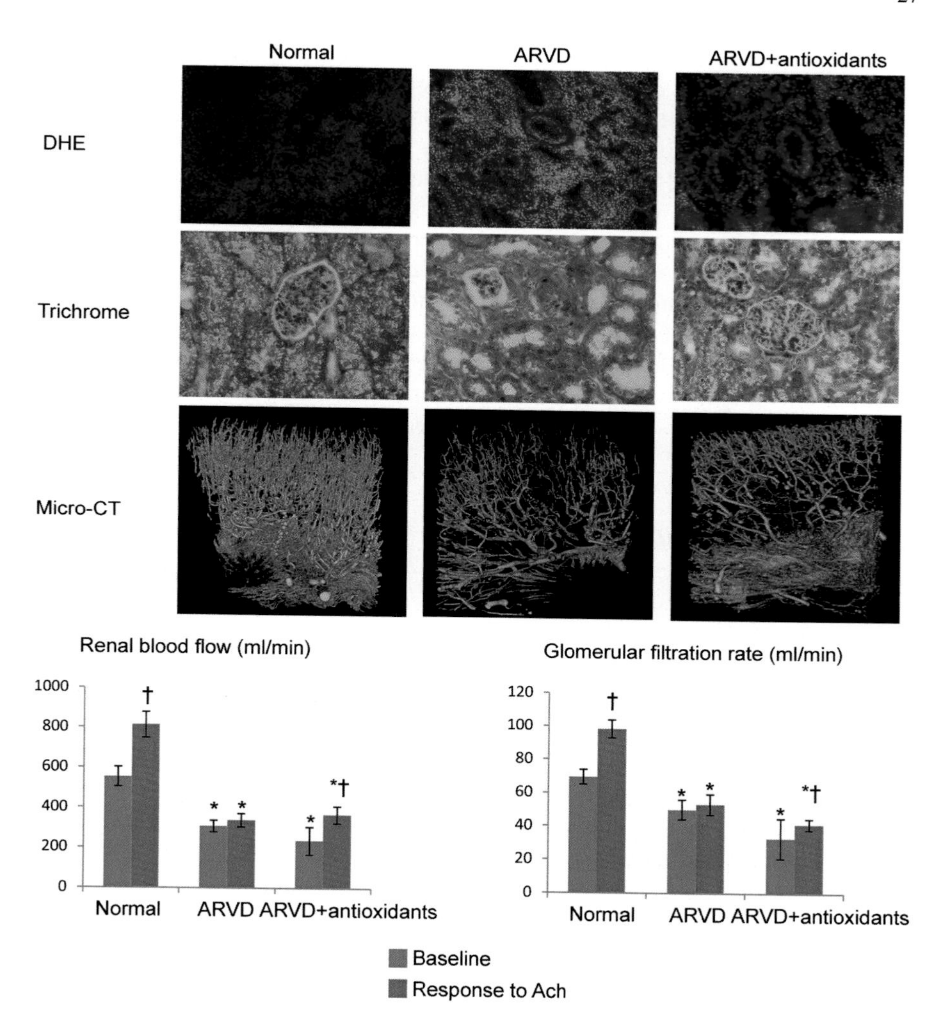

Fig. 2.3 Reactive oxygen species (ROS) and renal function, renovascular rarefaction, and renal fibrosis in porcine model of ARVD (with and without antioxidants). Representative images showing superoxide generation detected by dihydroethidium (DHE) staining and renal fibrosis by trichrome staining were increased in ARVD, and microcomputer tomograph (CT) images showing decreased microvascular density in ARVD, all were improved by antioxidants treatment. Furthermore, renal blood flow and glomerular filtration rate were both decreased in ARVD. Although antioxidants treatment didn't normalize renal function, renal blood flow response acetylcholine (Ach) was improved, suggest a role of ROS in endothelial dysfunction in ARVD

enalapril treatment, kidney length, renal blood flow, and glomerular filtration rate all decreased. Tubular cells were atrophic but not necrotic, accompanied by greatly reduced alkaline phosphatase in the tubular brush border and of acid phosphatase and peroxidase in lysosomes, while ATPase activity in the distal tubule segments remained unchanged. All observed changes were reversible within 2–3 weeks following removal of the clip. Importantly, ROS induced by growth factor withdrawal act upstream of caspases in the apoptotic pathway, and induce tubular cell apoptosis [45]. After I/R, apoptotic cells appeared principally in the tubular epithelial cells, but not in the interstitial cells, thereby indicating that ROS are particularly harmful in tubule cells [46]. Ang II-induced activation of mitochondrial Nox4 is an important endogenous source of ROS, and is related to cell survival [16]. Furthermore, aldosterone can also induces apoptosis via ROS-mediated, CHOP-dependent activation in renal tubular epithelial cells [47].

Using computed tomography imaging, we found that intratubular fluid concentration (ITC, an index of tubular fluid reabsorption) was decreased in both the proximal tubule and Henle's loop in ARVD compared to non-atherosclerotic renal artery stenosis. This might have represented early functional injury in these tubular segments [8].

Glomerular Function

GFR is determined by both renal blood flow and glomerular capillary hydrostatic pressure. When renal perfusion pressure falls distal to a stenotic site, the afferent arteriole dilates while the efferent arteriole constricts, resulting in increased filtration pressure. These intrarenal compensatory mechanisms maintain GFR and renal blood flow despite reductions in renal perfusion pressure by up to 40 % [48, 49]. However, the limited range of kidney autoregulation may restrict renal adaptation and may be impaired after prolonged hypoperfusion or hypertension. "Ischemic nephropathy" may ensue, which is defined as an obstruction of renal blood flow that leads to ischemia and excretory dysfunction. Global, focal, or segmental glomerulosclerosis have been found in both experimental [50] and clinical [2, 51] ARVD, but are usually late sequelae or associated with comorbidities. Furthermore, in patients with atherosclerotic nephropathy the severity of histopathological damage, including glomerulosclerosis and interstitial fibrosis, is an important determinant and predictor of renal functional outcome [2, 51].

Polymorphonuclear Neutrophls (PMN) play an important role in glomerular injury due their ability to release highly toxic ROS generated by the myeloperoxidase (MPO)-hydrogen peroxide-halide system. Rats infused with MPO showed marked glomerular injury to the endothelium and mesangial cells, as well as fusion of the glomerular epithelial foot processes and cationic MPO to glomerular basement membrane [52]. Recent studies showed that neutrophil-to-lymphocyte ratio was positively correlated with TNF-α in patients with end-stage renal disease. Importantly, ROS-induced renovascular constriction is inhibited by ROS scavengers [53], and antioxidants attenuate glomerulosclerosis in experimental ARVD [54]. Figure 2.3 shows GFR was significantly decreased in the stenotic kidney of pig ARVD, in association with increased DHE staining, indicating the important role of ROS in glomerular function.

Renal Fibrosis

In a patient cohort study, a significant correlation was observed between renal functional outcome and interstitial volume, suggesting interstitial fibrosis as an important determinant and predictor of renal functional outcome [51]. Histologic evaluation of biopsies from the stenotic kidneys of patients with subtotal atherosclerotic vascular occlusion demonstrate widespread tissue TGF-β staining associated with reduced blood flow. TGF-β expression was elevated despite relative preservation of tissue oxygenation in this ARVD cohort with relatively preserved renal function [55]. More severe decrements in blood flow were associated with higher severity of tissue fibrosis and tubular atrophy. In ARVD inflammatory cellular infiltrates, particularly CD68+ macrophages, were more prominent in both subtotal and total occlusion compared with biopsies from normal kidney donors, and their overall number correlated with TGF-β score [55]. Upregulation of TGF-β is also commonly observed in experimental renal artery stenosis [56, 57]. TGF-β expression is elevated in poststenotic kidneys during their remodeling process, and stimulates expression of MCP-1 in mesangial cells through pathways involving ROS generation, suggesting that this cascade promotes progressive renal disease. Indeed, abrogation of TGF-β/Smad3 signaling pathway confers protection against development of fibrosis and atrophy in murine renal artery stenosis [57]. Importantly, evidence indicates that ROS (H_2O_2) direct induce TGF-β1 synthesis and thereby increases extracellular matrix gene expression in cultured human mesangial cells [58]. These cellular responses may underlie the development and progression of renal fibrosis characterized by oxidative stress.

Expression of the pro-fibrotic mediators TIMP-1 and PAI-1 is upregulated in the stenotic kidney, and both localize to the tubular and interstitial compartments [59]. TIMP-1 inhibits extracellular matrix degradation, leading to accumulation of fibroblasts and collagen deposition, while PAI-1 plays an important role in Ang II-mediated hypertensive kidney and heart injury. Recent study showed that increased aortic collagen and elastin content in the early phase of renovascular hypertension, possibly as a result of increased vascular NAD(P)H oxidase activity and oxidative stress [60]. Therefore, treatment approaches targeted to block oxidative stress might prevent development of fibrosis and subsequent renal dysfunction.

Microvascular Architecture

Microvascular rarefaction in the stenotic kidney accelerates progression of renal injury in ARVD (Fig. 2.1). Microvessels (vessels <500 μm in diameter) are responsible for delivery of blood to the renal parenchyma and possess unique abilities to adapt to local metabolic demands, sustaining renal function in early stages of ARVD. Therefore, alterations in microvascular structure or function may lead to hypoperfused and hypo-oxygenated regions in the kidney, triggering matrix accumulation, interstitial fibrosis, and renal dysfunction. Importantly, reduced renal blood flow affects not only the number of microvessels in the stenotic kidney but also their structure and functionality. Indeed, Fig. 2.3 shows 3D microcomputed tomography images obtained from stenotic-kidney of ARVD pigs 10 weeks after induction of renal artery stenosis, illustrating significant impairment of the microvascular architecture and spatial density. Furthermore, decreased microvascular density was associated with enhanced oxidative stress. Importantly, antioxidants treatment partially restored microvascular density, underscore the role of ROS in microvascular remolding in ARVD [59].

Multiple pathways may contribute to microvascular damage in the stenotic kidney, including oxidative stress, apoptosis, inflammation, and fibrosis. In healthy men exposed to 12 h of sustained poikilocapnic hypoxia, 8-OHdG, advanced oxidation protein products, and vascular endothelial growth factor (VEGF) were increased in plasma and hypoxia inducible factor (HIF)-1 alpha mRNA was increased in leukocytes, suggest that hypoxia induces oxidative stress via an overgeneration of reactive oxygen species (ROS) [61]. Furthermore, Ang II can also increase HIF-1alpha in vascular smooth muscle cells to levels that are substantially more elevated than the hypoxic treatment [62]. However, hypoxia-induced ROS-dependent VEGF upregulation tend to increase vascular permeability, which were evidenced in human pulmonary artery endothelial cell (HPAEC) monolayers exposed to hypoxia, and treatment with antioxidants lowered the hypoxia-induced HPAEC monolayer permeability as well as the elevation of HIF-1alpha and VEGF [63]. Furthermore, downregulation of angiogenic factors like VEGF is often observed in the stenotic kidney [64] and is associated with decreased spatial density of cortical and medullary microvessels and enhanced oxidative stress, suggesting that ROS may impair HIF-VEGF pathway and contribute to vascular rarefaction. Indeed, ROS-dependent extracellular matrix depositing may limit microvessel sprouting, which also contribute to microvascular rarefaction in the stenotic kidney in ARVD.

Thus, ROS are involved in all major components of kidney injury in the stenotic kidney in ARVD, including endothelial dysfunction, glomerulosclerosis, microvascular rarefaction, and renal fibrosis.

Effects of ROS on the Non-stenotic Kidney

Vascular Function

The kidney damage in the non-stenotic kidney is mainly caused by hypertension. Typically, non-stenotic kidneys tend to have milder damage than post-stenotic kidneys [65]. In mice 2K1C model 11 weeks after renal artery stenosis, contralateral kidney showed minimal histopathological alterations [66]. Johnson et al. [67] suggested two stages of hypertension-induced kidney injury. In the first stage renal vasoconstriction dominates in the absence of altered renovascular structure, while in the second stage, renal vasoconstriction persists when the external stimuli are removed. In the first stage, as long as glomerular afferent arteriolar structures remain intact, renal autoregulation effectively prevents transmission of increases in systemic blood pressure to renal glomeruli or peritubular capillaries. This is accomplished by two intrarenal mechanisms: (1) the afferent arteriolar myogenic response, which is a reflex causing afferent arterioles to constrict in response to increased arterial pressure [68]. (2) Tubuloglomerular feedback, which alters afferent tone in response to altered Na and Cl concentrations in distal tubule as it passes the macula densa [69]. In stage two, patients develop salt sensitivity, renal arteriolar dysfunction, and impaired renal autoregulation. This stage constitutes a risk for developing microalbuminuria and progressive renal disease, which eventually results in end stage renal disease. Increased ROS induced by Ang II and elevated pressure also induces renal tissue hypoxia, since increased O_2 utilization is not compensated by increased O_2 delivery [70]. Sustained hypoxia induces fibrogenesis and tubular atrophy, which together with renovascular dysfunction, result in progressively diminishing kidney function. Ang II influences vascular tone via two distinctly different receptors; activation of AT1R causes vasoconstriction, whereas activation of AT2-receptors induces NO release and causes vasodilation [71]. Normally, AT1 receptors are more abundant, and constriction, therefore, dominates the vascular response to Ang II. Even a short-term exposure of vascular smooth muscle cells to Ang II results in contractile dysfunction [72]. These alterations result in hypertension and vascular and tubulointerstitial damage. Furthermore, renal damage and proteinuria are improved by inhibiting Ang II signaling in spontaneously hypertensive rats (SHR) but unaffected by similar blood pressure-lowering treatment with the calcium channel blocker amlodipine [73]. ROS may play important role in above pathophysiological alterations in the contralateral kidney. In experimental study, 30 min of unilateral renal ischemia (UI) resulted in gradual increase in contralateral kidney weight over time and increased superoxide formation. After UI, there was a significant increase in the number of NADPH oxidase 2 (Nox2)-expressing cells and the level of Nox2 expression in the contralateral kidney was observed. Treatments with superoxide dismutase (SOD) mimetic and apocynin (a putative NADPH oxidase inhibitor) inhibited UI-induced hypertrophy of CLK along with reduction in Nox2-positive cell, and superoxide formation [74]. Thus, renal mass reduction by UI may increase ROS formation in the contralateral kidney then subsequently structural and functional impairment.

Tubular Function

In a 2k1c model, unclipped kidney hypertrophy is considered to be a compensatory mechanism driven by growth factors. Hypertrophy of the renal tubular cells, especially those of the proximal tubule, accounts for the majority of the increase in kidney size that follows loss of renal mass in the stenotic kidney [75]. Epidermal growth factor (EGF) may drive the growth of proximal tubular cells in the contralateral rat kidney, as its expression increases 5 and 14 days post nephrectomy [76].

Furthermore, stretch caused an increase in EGF receptor phosphorylation and cytosolic to membrane translocation of the p47phox NAD(P)H oxidase subunit. Hydrogen peroxide also elicited contraction through EGF receptor phosphorylation [41]. In unilateral nephrectomized rats, secretion of IL-10 and TGF-beta by mesangial cells of the remaining kidneys contributes to the hypertrophy of tubular cells [77]. In a double transgenic mouse, activation of TGF-β signaling in the tubular epithelium alone was sufficient to cause AKI characterized by marked tubular cell apoptosis and necrosis, oxidative stress, dedifferentiation and regenerative cell proliferation, reduced renal function, and interstitial accumulation of inflammatory cells. This tubular injury was associated with mitochondrial-derived generation of ROS [78]. TGF-β1 treated renal tubular epithelial NRK-52E cells showed upregulated NAD(P)H oxidase subunit p67phox mRNA and significantly increase NAD(P)H oxidase-dependent intracellular ROS, MCP-1, and IL-6 expression [79]. Thus, TGF-β may be a mechanistic link between acute injury and chronic progression of kidney disease. Moreover, ROS-induced tubular cell apoptosis also contribute to tubular injury. Treatment with various doses of the aldehyde products of lipid peroxidation 4-hydroxy-2-hexenal (HHE) resulted in dose-dependent decreases of cell viability and increases of ROS. HHE decreased the expression of Bcl-2, while it increased that of Bax, which induced apoptosis [80].

Glomerular Function

Less attention has been paid to nephropathies and proteinuria in the contralateral kidney in ARVD. A clinical study in a small number of patients showed decreased GFR in the stenotic kidney and increased in the contralateral kidney compared with the right kidney of essential hypertension [81]. Focal segmental glomerulosclerosis and nephrosclerosis were found in contralateral kidney biopsies from patients with ARVD [82, 83]. Possibly, besides the increase in renal perfusion pressure, activation of RAAS might promote glomerular hyperfiltration through vasoconstriction of the efferent arterioles in the contralateral kidney. Focal segmental glomerulosclerosis-like lesion thus induced appeared to have caused massive proteinuria [84].

Importantly, increased perfusion pressure in the contralateral kidney induces oscillatory shear, which is associated with stretch of endothelial and vascular smooth muscle cells, which can directly activate NAD(P)H oxidase to generate ROS [85]. This effect may be amplified by activation of the RAAS. Increased oxidative stress in response to stretch contributes to activation of pro-inflammatory transcription factors, activation of growth-promoting mitogen-activated protein (MAP) kinases, upregulation of pro-fibrogenic mediators, and altered vascular tone, important processes contributing to the glomerulosclerosis in the contralateral kidney.

Renal Fibrosis

The contralateral kidney in a pig model of renal artery stenosis shows a modest increase in interstitial fibrosis compared with normal, and macrophage infiltration in these kidneys implicates inflammatory mechanisms [86]. ROS may mediate interstitial fibrosis in ARVD. A study uses type 2 diabetic (db/db) mice and in db/db transgenic (Tg) mice overexpressing rat catalase showed that Tg mice had significantly attenuated renal fibrosis and tubular apoptosis, indicating the pivotal role of ROS in the development of hypertension-induced renal fibrosis [87]. Furthermore, matrix metalloproteinases (MMPs) are important regulators for extracellular matrix remodeling and their expression were upregulated by increased formation of ROS [88]. Antioxidant approaches attenuated the increases in MMP-2 expression/activity and the vascular remodeling associated with 2K-1C hypertension [89].

We have mentioned that shear stress may induce ROS formation in renovascular hypertension, it may also be involved in the development of renal fibrosis. A recent study [90] reported that human renal tubular cells (HK-2) exposed to fluid shear stress promote human THP-1 monocytes toward the inflammatory M1-type macrophage. Fluid shear stress-injured HK-2 cells expressed and secreted early biomarkers of tubular damage such as kidney injury-molecule-1 and neutrophil gelatinase-associated lipocalin. Thus, changes in fluid shear stress should now also be considered as potential insults for tubular cells that initiate/perpetuate interstitial inflammation. In vitro experiments using proximal tubule epithelial cells demonstrated that pathological shear stresses induces TGF-beta1-SMAD pathway, which is mediated by Notch4 [91].

Microvascular Architecture

Unlike the striking microvascular remodeling in the stenotic kidney, vascular adaption in the contralateral kidney in unilateral renal artery stenosis may resemble that observed in other forms of hypertension, and includes intrarenal arterial hypertrophy. In SHR, the walls of the interlobar, arcuate, and interlobular arteries appear to be hypertrophied in both the "pre-hypertensive" phase and in established hypertension, which is not reversible by chronic angiotensin converting enzyme inhibition. It is not easy to document changes in wall dimensions of intrarenal arteries during the development of human hypertension, but renal hemodynamic abnormalities currently attributed to renal vasoconstriction in early human hypertension are also compatible with renal arterial hypertrophy. These abnormalities include increased resting renal vascular resistance and augmented renal vascular resistance responses to vasoconstrictor agents. Hypertrophy of the renal vasculature to increase pre-glomerular resistance may have dual effects: increased renal vascular resistance, and effects on renal hemodynamics distally in a manner similar to narrowing of the main renal artery [92]. Intrarenal arterial hypertrophy may be induced by Ang II mediated pleiotropic vascular effects through NAD(P)H oxidase-derived ROS. Furthermore oscillatory shear stress induced by hypertension is linked to increased ROS production with consequent oxidative damage. Induction of these signaling cascades leads to vascular smooth muscle cell growth and migration, expression of pro-inflammatory mediators, and modification of extracellular matrix [85].

Although studies on the pathophysiology of the non-stenotic kidney are much less intense, ROS clearly played important role in vascular and tubular injury. Preserve renal function in the non-stenotic kidney may be more important compared with restore damaged stenotic kidney.

Effects of ROS on Blood Pressure

Evidences of oxidative stress have been found in most experimental models of hypertension. Mice with ROS-generating enzyme deficiencies have lower blood pressure compared with wild-type counterparts [93], and fail to induce hypertension after Ang II infusion [94]. Thus ROS are critical in Ang II-induced hypertension, at least in animal models. However, most human studies showing only indirect associations between ROS and blood pressure. ROS participate in several redox-sensitive pathways involving the development of hypertension, particularly in the vasculature, kidney, and central nerves system. Superoxide anion and H_2O_2, act as second messengers in a highly regulated manner, stimulate MAP kinases, tyrosine kinases, Rho kinase, and transcription factors (NF-κB, activator protein -1, and HIF-1), inactivate protein tyrosine phosphatases, and proinflammatory gene expression and activity [95, 96]. In the vasculature, these alterations lead to endothelial dysfunction, reduced vasodilation, enhanced contraction, and structural remodeling, which in concert increase peripheral resistance and elevate blood pressure. In the kidney, activation of redox-sensitive pathways is associated with glomerular damage, proteinuria, sodium and water retention, and nephron loss, all of which are important in the development of hypertension. In the central nerves system, ROS produced by NAD(P)H oxidase in the hypothalamic and circumventricular organs are implicated in central control of hypertension [97, 98], in part through sympathetic outflow. Although oxidative stress may play a role in the pathophysiology of hypertension and associated target-organ damage, it is likely not the sole cause of blood pressure elevation. Ang II is a potent inducer of oxidative stress and Ang II-dependent hypertension is particularly sensitive to Nox-derived ROS [99]. Despite abundant experimental data supporting an etiological role for oxidative stress in the pathogenesis of hypertension, there is no confirmation that oxidative stress is a primary cause of hypertension in humans. Nevertheless, there is evidence that ROS bioavailability is increased in patients with essential hypertension, renovascular hypertension, malignant hypertension, salt-sensitive hypertension, cyclosporine-induced hypertension, and pre-eclampsia. Moreover, a link between oxidative stress and cardiovascular injury and hypertension associated target organ damage has been suggested [100]. Most clinical study findings are based on increased levels of plasma thiobarbituric acid-reactive substances and 8-epi-isoprostanes, biomarkers of lipid peroxidation, and oxidative stress.

Interventions to Decrease ROS in ARVD

Antioxidants

In experimental settings, natural or synthesized antioxidants have been applied for protection of kidney in renovascular disease [89, 101]. SOD mimic tempol normalized blood pressure in SHR in both short term and long term administration [102, 103], suggests a role for oxygen radicals in the maintenance of hypertension in SHR. Long-term feeding of blueberry-enriched diet lowered blood pressure, preserved renal hemodynamics, and improved redox status in kidneys of hypertensive rats and concomitantly demonstrated the potential to delay or attenuate development of hypertension-induced renal injury [104]. We supplemented antioxidant vitamins in a swine model of ARVD [54]. Kidney hemodynamics and function were quantified after 12 weeks of experimental ARVD (simulated by concurrent hypercholesterolemia and renal artery stenosis) with or without daily supplemented with antioxidant vitamins C (1 g) and E (100 IU/kg). We found that basal RBF and GFR were decreased in the stenotic kidney of ARVD pigs regardless of antioxidant supplement, but RBF and GFR response to acetylcholine that was blunted in ARVD significantly improved in vitamin-treated pigs. Vitamins also showed increased renal expression of endothelial nitric oxide synthase and decreased expression of NAD(P)H-oxidase, nitrotyrosine, inducible-nitric oxide synthase, and NF-κB, suggesting decreased superoxide abundance and inflammation. Furthermore, decreased expression of pro-fibrotic factors in vitamin-treated pigs was accompanied by augmented expression of extracellular (matrix metalloproteinase–2) and intracellular (ubiquitin) protein degradation systems, resulting in significantly attenuated glomerulosclerosis and renal fibrosis. Thus, chronic antioxidant intervention in early experimental ARVD improved renal endothelial function, enhanced tissue remodeling, and decreased structural injury. In human studies, the variable results of antioxidant vitamin intervention observed in clinical studies [105–107] are likely related to differences in study population, the duration, dose, and type of supplements, as well as outcome measures. The feasibility of applying antioxidant strategies for preserving the atherosclerotic and ischemic kidney still needs further investigations.

Although there may be some debate about whether it is NAD(P)H oxidase or other oxygenases responsible for kidney damage in hypertension, it is clear that reduction in NAD(P)H oxidase activity, and hence ROS, can ameliorate glomerular filtration barrier injury. Novel approaches to block NAD (P)H oxidase activity including gp91 ds-tat, siRNA [108], and monoclonal antibodies are being tried and hold great promise for not only hypertension but also for other disease processes with elevated oxidative stress.

Mitochondria are the important source of ROS. Recently, we have applied Bendavia, a novel tetrapeptide that inhibits mitochondrial permeability transition pore opening, in swine ARVD, and demonstrated that it protects the stenotic kidney by reducing oxidative stress and apoptosis [109, 110].

RAAS Inhibitors

Medical therapy is the cornerstone of treatment for ARVD, either alone or in combination with revascularization. The goals are reducing anatomic progression of the lesion and the impact of its consequences, improving blood pressure control, preserving renal function, slowing the rate of progression to ESRD, and reducing cardiovascular events. Inhibition of the RAAS with ACE inhibitors and angiotensin receptor blockers (ARBs) affords cardiovascular morbidity and mortality benefits in patients with cardiovascular risk [111]. Furthermore, RAAS inhibition reduces proteinuria and may delay ESRD in patients with renal dysfunction [112]. ACEI and ARB are effective in hypertension treatment in the presence of ARVD. There is some evidence that ACE inhibitors have the most potent blood pressure-lowering effect in these patients [113]. Furthermore, in nonrandomized studies of patients with ARVD, ACE inhibitor use was independently associated with survival benefit regardless of whether patients underwent revascularization [114, 115]. Similarly, in a cohort study [116] of 3570 patients with ARVD, ACE inhibitor use was associated with a lower risk for the primary composite outcome of death, and dialysis initiation [117]. Despite the potential benefits, use of RAAS inhibitors need to be monitored closely in patients with ARVD, as precipitous declines in GFR may occur. Caution should be taken in patients with older age, higher baseline creatinine (>2.0) and/or lower eGFR (<35), although a prospective cohort study reported that eGFR improved following discontinuation of RAAS blockade [118].

The mechanisms for improvement in renal function and/or proteinuria by RAAS inhibitors are not only related to their blood pressure lowering effects, but also depend on their anti-oxidant, anti-inflammatory, and anti-fibrotic effects. In transgenic rat models and genetic models of hypertension and glomerular filtration barrier injury, the ARBs/ACE inhibitors valsartan, irbesartan, and losartan reduced blood pressure, and the tissue benefits were beyond that anticipated by BP control alone including amelioration of proteinuria, slit diaphragm widening, podocyte effacement, decrease in number of slit pores and basement membrane widening. Furthermore, these treatments decreased perivascular fibrosis and oxidative stress (3-nitrotyrosine in glomeruli, 8-OHdG in urine, NAD(P)H oxidase activity, enzyme subunits by western blots and immunohistochemistry). No specific clinical trial has been designed to test the anti-oxidant property of AECI/ARB. However, in 53 nondialyzed hypertensive CKD patients taking ACE inhibitor or calcium blocker, serum and urinary 8-OHdG were determined. In comparison to a calcium channel blocker, an ACE inhibitor seems much more protective against oxidative DNA damage in hypertensive patients with different stages of CKD [119].

Percutaneous Transluminal Renal Angioplasty (PTRA)

The rationale for renal artery revascularization is to relieve obstruction to flow and downstream ischemia, as well as to reduce activation of the RAAS and the subsequent cascade of pathophysiological processes. Although early nonrandomized studies showed some benefits of revascularization in ARVD [120, 121], recent large clinical trials revealed that renal-artery stenting did not confer a significant benefit with respect to the prevention of clinical events when added to comprehensive, multifactorial medical therapy in patients with ARVD and hypertension or chronic kidney disease [122]. In experimental ARVD, we found that 4 weeks after renal artery stenting blood pressure was normalized. However, GFR and RBF remained unchanged. Microvascular rarefaction was unaltered after revascularization, and the spatial density of outer cortical microvessels correlated with residual GFR. Interstitial fibrosis and altered expression of proangiogenic and profibrotic factors persisted after stenting. Microvascular loss and fibrosis in swine ARVD might account for persistent renal dysfunction after revascularization and underscore the need to assess renal parenchymal disease before revascularization [123]. Interestingly, Ziakka et al. found that oxidative stress was the strongest predictive factor for serum creatinine increase in these patients who failed to improve renal function after revascularization [121]. We also implicated inflammation in the inability of PTRA to fully reverse renal damage in the post-stenotic kidney [124, 125]. On the other hand, renal angioplasty decreased plasma renin activity, plasma AngII concentration, and serum MDA-LDL concentration and urinary 8-OHdG excretion in patients with renovascular hypertension, and forearm blood flow response to acetylcholine was enhanced, suggest that PTCA may attenuate oxidative stress in a clinical setting [42].

Stem Cells

Mesenchymal stem cells (MSC) confer renal protection through paracrine/endocrine effects, and their anti-inflammatory and immune-modulatory properties target multiple cascades in the mechanisms of ischemic kidney in ARVD. As discussed previously, clinical trials have not identified major benefits for PTRA [126], likely due to lingering kidney tissue damage. To improve its efficiency, we replenished MSC as an adjunct to experimental PTRA in ARVD pigs [127]. PTRA was performed 6 weeks after renal artery stenosis, with adjunct delivery of adipose tissue-derived-MSC (10×10^6 cells). Four weeks after successful PTRA, mean arterial pressure fell to similar levels in all revascularized pigs. MSC restored stenotic-kidney GFR and RBF, which remained low after PTRA alone. Interstitial fibrosis, inflammation, microvascular rarefaction, and oxidative stress were also attenuated to a greater degree in PTRA + MSC-treated pigs. This study suggested a novel therapeutic potential for MSC in restoring renal function and blunting structural remodeling when combined with PTRA in ARVD.

The mechanisms by which MSC achieve renal cellular repair are multifactorial. MSC ameliorate I/R-induced renal dysfunction by improving the antioxidant enzymes superoxide dismutase and glutathione peroxidase levels [128] and through enhancing NAD(P)H quinone oxidoreductase-1 and HO-1 activities, two indicators of anti-oxidative capacity [129]. Furthermore, MSC may release growth factors or anti-inflammatory cytokines to the injury site. MSC release microparticles carrying anti-inflammatory cytokines and growth factors that promote kidney repair by their internalization in tubular or other cells. All these actions tone down intra-renal inflammation and oxidative stress, and allow for vascular regeneration. Moreover, anti-apoptotic effects of MSC [130] can prevent cell loss.

Summary

In summary, ROS played pivotal role in the pathophysiology of ARVD, in both acute and chronic renal injuries. Interventions with antioxidants have different effects on blood pressure. SOD mimic tempol was able to normalize blood pressure in SHR, reflects that ROS is the major factor in the maintenance of hypertension in this model. While antioxidant vitamins have potential to improve renal function in experimental ARVD but failed to normalize blood pressure, suggest that ROS is not the only mechanism in renovascular hypertension. Although we have gained plenty of knowledge in ROS formation and their signaling in different diseases, the basic physiological role of ROS is still not clear. We may still a few steps away from fully understand the entire picture of ROS. Nevertheless, blockade of RAAS using ACEI/ARB have showed potential of attenuation of oxidative stress while improving clinical outcome in ARVD. Stem cells combine with PTRA has promising results in experimental ARVD and translational studies are urgently needed in this area.

References

1. Hansen KJ, Edwards MS, Craven TE, Cherr GS, Jackson SA, Appel RG, et al. Prevalence of renovascular disease in the elderly: a population-based study. J Vasc Surg. 2002;36:443–51.
2. Keddis MT, Garovic VD, Bailey KR, Wood CM, Raissian Y, Grande JP. Ischaemic nephropathy secondary to atherosclerotic renal artery stenosis: clinical and histopathological correlates. Nephrol Dial Transplant. 2010;25:3615–22.
3. Sarnak MJ, Levey AS, Schoolwerth AC, Coresh J, Culleton B, Hamm LL, et al. Kidney disease as a risk factor for development of cardiovascular disease: a statement from the American Heart Association Councils on Kidney in Cardiovascular Disease, High Blood Pressure Research, Clinical Cardiology, and Epidemiology and Prevention. Circulation. 2003;108:2154–69.
4. Goldblatt H, Lynch J, Hanzal RF, Summerville WW. Studies on experimental hypertension : I. The production of persistent elevation of systolic blood pressure by means of renal ischemia. J Exp Med. 1934;59:347–79.
5. Lerman LO, Nath KA, Rodriguez-Porcel M, Krier JD, Schwartz RS, Napoli C, et al. Increased oxidative stress in experimental renovascular hypertension. Hypertension. 2001;37:541–6.
6. Lerman LO, Schwartz RS, Grande JP, Sheedy PF, Romero JC. Noninvasive evaluation of a novel swine model of renal artery stenosis. J Am Soc Nephrol. 1999;10:1455–65.
7. Vasilev T, Kiprov D, Puchlev A, Todorova L. Plasma renin activity in patients with renovascular hypertension. Cor Vasa. 1978;20:35–43.
8. Chade AR, Rodriguez-Porcel M, Grande JP, Krier JD, Lerman A, Romero JC, et al. Distinct renal injury in early atherosclerosis and renovascular disease. Circulation. 2002;106:1165–71.

9. Hartono SP, Knudsen BE, Zubair AS, Nath KA, Textor SJ, Lerman LO, et al. Redox signaling is an early event in the pathogenesis of renovascular hypertension. Int J Mol Sci. 2013; 14:18640–56.
10. Marks LS, Maxwell MH. Tigerstedt and the discovery of renin. An historical note. Hypertension. 1979;1:384–8.
11. Braun-Menendez E, Page IH. Suggested revision of nomenclature—angiotensin. Science. 1958;127:242.
12. Skeggs Jr LT, Kahn JR, Shumway NP. The purification of hypertensin II. J Exp Med. 1956; 103:301–7.
13. Fyhrquist F, Saijonmaa O. Renin-angiotensin system revisited. J Intern Med. 2008;264: 224–36.
14. Aroor AR, Demarco VG, Jia G, Sun Z, Nistala R, Meininger GA, et al. The role of tissue renin-angiotensin-aldosterone system in the development of endothelial dysfunction and arterial stiffness. Front Endocrinol. 2013;4:161.
15. Ikemoto F, Song GB, Tominaga M, Kanayama Y, Yamamoto K. Angiotensin-converting enzyme in the rat kidney. Activity, distribution, and response to angiotensin-converting enzyme inhibitors. Nephron. 1990;55 Suppl 1:3–9.
16. Kim SM, Kim YG, Jeong KH, Lee SH, Lee TW, Ihm CG, et al. Angiotensin II-induced mitochondrial Nox4 is a major endogenous source of oxidative stress in kidney tubular cells. PLoS One. 2012;7, e39739.
17. Zhou A, Carrell RW, Murphy MP, Wei Z, Yan Y, Stanley PL, et al. A redox switch in angiotensinogen modulates angiotensin release. Nature. 2010;468:108–11.
18. Ozawa Y, Kobori H, Suzaki Y, Navar LG. Sustained renal interstitial macrophage infiltration following chronic angiotensin II infusions. Am J Physiol Renal Physiol. 2007;292:F330–9.
19. Li L, Huang L, Sung SS, Vergis AL, Rosin DL, Rose Jr CE, et al. The chemokine receptors CCR2 and CX3CR1 mediate monocyte/macrophage trafficking in kidney ischemia-reperfusion injury. Kidney Int. 2008;74:1526–37.
20. Wynn T. Cellular and molecular mechanisms of fibrosis. J Pathol. 2008;214:199–210.
21. Li P, Garcia GE, Xia Y, Wu W, Gersch C, Park PW, et al. Blocking of monocyte chemoattractant protein-1 during tubulointerstitial nephritis resulted in delayed neutrophil clearance. Am J Pathol. 2005;167:637–49.
22. Alvarez A, Cerda-Nicolas M, Naim Abu Nabah Y, Mata M, Issekutz AC, Panes J, et al. Direct evidence of leukocyte adhesion in arterioles by angiotensin II. Blood. 2004;104:402–8.
23. Phillips MI, Kagiyama S. Angiotensin II as a pro-inflammatory mediator. Curr Opin Investig Drugs. 2002;3:569–77.
24. Kranzhofer R, Schmidt J, Pfeiffer CA, Hagl S, Libby P, Kubler W. Angiotensin induces inflammatory activation of human vascular smooth muscle cells. Arterioscler Thromb Vasc Biol. 1999;19:1623–9.
25. Ni W, Kitamoto S, Ishibashi M, Usui M, Inoue S, Hiasa K, et al. Monocyte chemoattractant protein-1 is an essential inflammatory mediator in angiotensin II-induced progression of established atherosclerosis in hypercholesterolemic mice. Arterioscler Thromb Vasc Biol. 2004;24:534–9.
26. Moreno-Manzano V, Ishikawa Y, Lucio-Cazana J, Kitamura M. Selective involvement of superoxide anion, but not downstream compounds hydrogen peroxide and peroxynitrite, in tumor necrosis factor-alpha-induced apoptosis of rat mesangial cells. J Biol Chem. 2000;275:12684–91.
27. Rajamohan SB, Raghuraman G, Prabhakar NR, Kumar GK. NADPH oxidase-derived H(2) O(2) contributes to angiotensin II-induced aldosterone synthesis in human and rat adrenal cortical cells. Antioxid Redox Signal. 2012;17:445–59.
28. Stouffer GA, Pathak A, Rojas M. Unilateral renal artery stenosis causes a chronic vascular inflammatory response in ApoE–/– mice. Trans Am Clin Climatol Assoc. 2010;121:252–64. 264-6.
29. Schleicher E, Friess U. Oxidative stress, AGE, and atherosclerosis. Kidney Int Suppl. 2007; S17–26.

30. Basta G, Lazzerini G, Del Turco S, Ratto GM, Schmidt AM, De Caterina R. At least 2 distinct pathways generating reactive oxygen species mediate vascular cell adhesion molecule-1 induction by advanced glycation end products. Arterioscler Thromb Vasc Biol. 2005;25: 1401–7.
31. Gao X, Zhang H, Schmidt AM, Zhang C. AGE/RAGE produces endothelial dysfunction in coronary arterioles in type 2 diabetic mice. Am J Physiol Heart Circ Physiol. 2008;295: H491–8.
32. Johnson KJ, Weinberg JM. Postischemic renal injury due to oxygen radicals. Curr Opin Nephrol Hypertens. 1993;2:625–35.
33. Kaminski KA, Bonda TA, Korecki J, Musial WJ. Oxidative stress and neutrophil activation—the two keystones of ischemia/reperfusion injury. Int J Cardiol. 2002;86:41–59.
34. Jang HR, Rabb H. The innate immune response in ischemic acute kidney injury. Clin Immunol. 2009;130:41–50.
35. Liu H, Liu S, Li Y, Wang X, Xue W, Ge G, et al. The role of SDF-1-CXCR4/CXCR7 axis in the therapeutic effects of hypoxia-preconditioned mesenchymal stem cells for renal ischemia/reperfusion injury. PLoS One. 2012;7, e34608.
36. Wei Q, Bhatt K, He HZ, Mi QS, Haase VH, Dong Z. Targeted deletion of Dicer from proximal tubules protects against renal ischemia-reperfusion injury. J Am Soc Nephrol. 2010;21:756–61.
37. Cantaluppi V, Gatti S, Medica D, Figliolini F, Bruno S, Deregibus MC, et al. Microvesicles derived from endothelial progenitor cells protect the kidney from ischemia-reperfusion injury by microRNA-dependent reprogramming of resident renal cells. Kidney Int. 2012;82: 412–27.
38. Lubas A, Zelichowski G, Prochnicka A, Wisniewska M, Wankowicz Z. Renal autoregulation in medical therapy of renovascular hypertension. Arch Med Sci. 2010;6:912–8.
39. Textor SC, Lerman L. Renovascular hypertension and ischemic nephropathy. Am J Hypertens. 2010;23:1159–69.
40. Goldsmith SR. Interactions between the sympathetic nervous system and the RAAS in heart failure. Curr Heart Fail Rep. 2004;1:45–50.
41. Oeckler RA, Kaminski PM, Wolin MS. Stretch enhances contraction of bovine coronary arteries via an NAD(P)H oxidase-mediated activation of the extracellular signal-regulated kinase mitogen-activated protein kinase cascade. Circ Res. 2003;92:23–31.
42. Higashi Y, Sasaki S, Nakagawa K, Matsuura H, Oshima T, Chayama K. Endothelial function and oxidative stress in renovascular hypertension. N Engl J Med. 2002;346:1954–62.
43. Gobe GC, Axelsen RA, Searle JW. Cellular events in experimental unilateral ischemic renal atrophy and in regeneration after contralateral nephrectomy. Lab Invest. 1990;63:770–9.
44. Grone HJ, Warnecke E, Olbricht CJ. Characteristics of renal tubular atrophy in experimental renovascular hypertension: a model of kidney hibernation. Nephron. 1996;72:243–52.
45. Lieberthal W, Triaca V, Koh JS, Pagano PJ, Levine JS. Role of superoxide in apoptosis induced by growth factor withdrawal. Am J Physiol. 1998;275:F691–702.
46. Kim J, Jung KJ, Park KM. Reactive oxygen species differently regulate renal tubular epithelial and interstitial cell proliferation after ischemia and reperfusion injury. Am J Physiol Renal Physiol. 2010;298:F1118–29.
47. Ding W, Yang L, Zhang M, Gu Y. Reactive oxygen species-mediated endoplasmic reticulum stress contributes to aldosterone-induced apoptosis in tubular epithelial cells. Biochem Biophys Res Commun. 2012;418:451–6.
48. Oien AH, Aukland K. A mathematical analysis of the myogenic hypothesis with special reference to autoregulation of renal blood flow. Circ Res. 1983;52:241–52.
49. Hope A, Clausen G, Rosivall L. Total and local renal blood flow and filtration in the rat during reduced renal arterial blood pressure. Acta Physiol Scand. 1981;113:455–63.
50. Chade AR, Zhu XY, Grande JP, Krier JD, Lerman A, Lerman LO. Simvastatin abates development of renal fibrosis in experimental renovascular disease. J Hypertens. 2008;26: 1651–60.

51. Wright JR, Duggal A, Thomas R, Reeve R, Roberts IS, Kalra PA. Clinicopathological correlation in biopsy-proven atherosclerotic nephropathy: implications for renal functional outcome in atherosclerotic renovascular disease. Nephrol Dial Transplant. 2001;16:765–70.
52. Johnson RJ, Couser WG, Chi EY, Adler S, Klebanoff SJ. New mechanism for glomerular injury. Myeloperoxidase-hydrogen peroxide-halide system. J Clin Invest. 1987;79:1379–87.
53. Rahman MM, Varghese Z, Fuller BJ, Moorhead JF. Renal vasoconstriction induced by oxidized LDL is inhibited by scavengers of reactive oxygen species and L-arginine. Clin Nephrol. 1999;51:98–107.
54. Chade AR, Rodriguez-Porcel M, Herrmann J, Zhu X, Grande JP, Napoli C, et al. Antioxidant intervention blunts renal injury in experimental renovascular disease. J Am Soc Nephrol. 2004;15:958–66.
55. Gloviczki ML, Keddis MT, Garovic VD, Friedman H, Herrmann S, McKusick MA, et al. TGF expression and macrophage accumulation in atherosclerotic renal artery stenosis. Clin J Am Soc Nephrol. 2013;8:546–53.
56. Urbieta-Caceres VH, Zhu XY, Jordan KL, Tang H, Textor K, Lerman A, et al. Selective improvement in renal function preserved remote myocardial microvascular integrity and architecture in experimental renovascular disease. Atherosclerosis. 2012;221:350–8.
57. Warner GM, Cheng J, Knudsen BE, Gray CE, Deibel A, Juskewitch JE, et al. Genetic deficiency of Smad3 protects the kidneys from atrophy and interstitial fibrosis in 2K1C hypertension. Am J Physiol Renal Physiol. 2012;302:F1455–64.
58. Iglesias De La Cruz MC, Ruiz-Torres P, Alcami J, Diez-Marques L, Ortega-Velazquez R, Chen S, et al. Hydrogen peroxide increases extracellular matrix mRNA through TGF-beta in human mesangial cells. Kidney Int. 2001;59:87–95.
59. Chade AR, Rodriguez-Porcel M, Grande JP, Zhu X, Sica V, Napoli C, et al. Mechanisms of renal structural alterations in combined hypercholesterolemia and renal artery stenosis. Arterioscler Thromb Vasc Biol. 2003;23:1295–301.
60. Ceron CS, Rizzi E, Guimaraes DA, Martins-Oliveira A, Cau SB, Ramos J, et al. Time course involvement of matrix metalloproteinases in the vascular alterations of renovascular hypertension. Matrix Biol. 2012;31:261–70.
61. Pialoux V, Mounier R, Brown AD, Steinback CD, Rawling JM, Poulin MJ. Relationship between oxidative stress and HIF-1 alpha mRNA during sustained hypoxia in humans. Free Radic Biol Med. 2009;46:321–6.
62. Richard DE, Berra E, Pouyssegur J. Nonhypoxic pathway mediates the induction of hypoxia-inducible factor 1alpha in vascular smooth muscle cells. J Biol Chem. 2000;275:26765–71.
63. Irwin DC, McCord JM, Nozik-Grayck E, Beckly G, Foreman B, Sullivan T, et al. A potential role for reactive oxygen species and the HIF-1alpha-VEGF pathway in hypoxia-induced pulmonary vascular leak. Free Radic Biol Med. 2009;47:55–61.
64. Zhu XY, Chade AR, Rodriguez-Porcel M, Bentley MD, Ritman EL, Lerman A, et al. Cortical microvascular remodeling in the stenotic kidney: role of increased oxidative stress. Arterioscler Thromb Vasc Biol. 2004;24:1854–9.
65. Fujii H, Nakamura S, Kuroda S, Yoshihara F, Nakahama H, Inenaga T, et al. Relationship between renal artery stenosis and intrarenal damage in autopsy subjects with stroke. Nephrol Dial Transplant. 2006;21:113–9.
66. Cheng J, Zhou W, Warner GM, Knudsen BE, Garovic VD, Gray CE, et al. Temporal analysis of signaling pathways activated in a murine model of two-kidney, one-clip hypertension. Am J Physiol Renal Physiol. 2009;297:F1055–68.
67. Johnson RJ, Segal MS, Srinivas T, Ejaz A, Mu W, Roncal C, et al. Essential hypertension, progressive renal disease, and uric acid: a pathogenetic link? J Am Soc Nephrol. 2005;16:1909–19.
68. Hayashi K, Epstein M, Saruta T. Altered myogenic responsiveness of the renal microvasculature in experimental hypertension. J Hypertens. 1996;14:1387–401.
69. Liu R, Carretero OA, Ren Y, Garvin JL. Increased intracellular pH at the macula densa activates nNOS during tubuloglomerular feedback. Kidney Int. 2005;67:1837–43.

70. Welch WJ, Mendonca M, Aslam S, Wilcox CS. Roles of oxidative stress and AT1 receptors in renal hemodynamics and oxygenation in the postclipped 2 K,1C kidney. Hypertension. 2003;41:692–6.
71. Endo Y, Arima S, Yaoita H, Tsunoda K, Omata K, Ito S. Vasodilation mediated by angiotensin II type 2 receptor is impaired in afferent arterioles of young spontaneously hypertensive rats. J Vasc Res. 1998;35:421–7.
72. Bochkov VN, Tkachuk VA, Hahn AW, Bernhardt J, Buhler FR, Resink TJ. Concerted effects of lipoproteins and angiotensin II on signal transduction processes in vascular smooth muscle cells. Arterioscler Thromb. 1993;13:1261–9.
73. Izuhara Y, Nangaku M, Inagi R, Tominaga N, Aizawa T, Kurokawa K, et al. Renoprotective properties of angiotensin receptor blockers beyond blood pressure lowering. J Am Soc Nephrol. 2005;16:3631–41.
74. Jang HS, Kim JI, Kim J, Na YK, Park JW, Park KM. Bone marrow derived cells and reactive oxygen species in hypertrophy of contralateral kidney of transient unilateral renal ischemia-induced mouse. Free Radic Res. 2012;46:903–11.
75. Moller JC. Proximal tubules in long-term compensatory renal growth. Quantitative light- and electron-microscopic analyses. APMIS Suppl. 1988;4:82–6.
76. Miller SB, Rogers SA, Estes CE, Hammerman MR. Increased distal nephron EGF content and altered distribution of peptide in compensatory renal hypertrophy. Am J Physiol. 1992;262:F1032–8.
77. Sinuani I, Averbukh Z, Gitelman I, Rapoport MJ, Sandbank J, Albeck M, et al. Mesangial cells initiate compensatory renal tubular hypertrophy via IL-10-induced TGF-beta secretion: effect of the immunomodulator AS101 on this process. Am J Physiol Renal Physiol. 2006;291:F384–94.
78. Gentle ME, Shi S, Daehn I, Zhang T, Qi H, Yu L, et al. Epithelial cell TGFbeta signaling induces acute tubular injury and interstitial inflammation. J Am Soc Nephrol. 2013;24: 787–99.
79. Zhang H, Jiang Z, Chang J, Li X, Zhu H, Lan HY, et al. Role of NAD(P)H oxidase in transforming growth factor-beta1-induced monocyte chemoattractant protein-1 and interleukin-6 expression in rat renal tubular epithelial cells. Nephrology. 2009;14:302–10.
80. Bae EH, Cho S, Joo SY, Ma SK, Kim SH, Lee J, et al. 4-Hydroxy-2-hexenal-induced apoptosis in human renal proximal tubular epithelial cells. Nephrol Dial Transplant. 2011;26: 3866–73.
81. Kimura G, London GM, Safar ME, Kuramochi M, Omae T. Glomerular hypertension in renovascular hypertensive patients. Kidney Int. 1991;39:966–72.
82. Alchi B, Shirasaki A, Narita I, Nishi S, Ueno M, Saeki T, et al. Renovascular hypertension: a unique cause of unilateral focal segmental glomerulosclerosis. Hypertens Res. 2006;29: 203–7.
83. Bhowmik D, Dash SC, Jain D, Agarwal SK, Tiwari SC, Dinda AK. Renal artery stenosis and focal segmental glomerulosclerosis in the contralateral kidney. Nephrol Dial Transplant. 1998;13:1562–4.
84. Ubara Y, Hara S, Katori H, Yamada A, Morii H. Renovascular hypertension may cause nephrotic range proteinuria and focal glomerulosclerosis in contralateral kidney. Clin Nephrol. 1997;48:220–3.
85. Paravicini TM, Touyz RM. Redox signaling in hypertension. Cardiovasc Res. 2006;71: 247–58.
86. Zhu XY, Chade AR, Krier JD, Daghini E, Lavi R, Guglielmotti A, et al. The chemokine monocyte chemoattractant protein-1 contributes to renal dysfunction in swine renovascular hypertension. J Hypertens. 2009;27:2063–73.
87. Brezniceanu ML, Liu F, Wei CC, Chenier I, Godin N, Zhang SL, et al. Attenuation of interstitial fibrosis and tubular apoptosis in db/db transgenic mice overexpressing catalase in renal proximal tubular cells. Diabetes. 2008;57:451–9.
88. Rizzi E, Guimaraes DA, Ceron CS, Prado CM, Pinheiro LC, Martins-Oliveira A, et al. Beta-adrenergic blockers exert antioxidant effects, reduce matrix metalloproteinase activity, and

improve renovascular hypertension-induced cardiac hypertrophy. Free Radic Biol Med. 2014;73C:308–17.
89. Castro MM, Rizzi E, Rodrigues GJ, Ceron CS, Bendhack LM, Gerlach RF, et al. Antioxidant treatment reduces matrix metalloproteinase-2-induced vascular changes in renovascular hypertension. Free Radic Biol Med. 2009;46:1298–307.
90. Miravete M, Dissard R, Klein J, Gonzalez J, Caubet C, Pecher C, et al. Renal tubular fluid shear stress facilitates monocyte activation toward inflammatory macrophages. Am J Physiol Renal Physiol. 2012;302:F1409–17.
91. Grabias BM, Konstantopoulos K. Notch4-dependent antagonism of canonical TGF-beta1 signaling defines unique temporal fluctuations of SMAD3 activity in sheared proximal tubular epithelial cells. Am J Physiol Renal Physiol. 2013;305:F123–33.
92. Anderson WP, Kett MM, Evans RG, Alcorn D. Pre-glomerular structural changes in the renal vasculature in hypertension. Blood Press Suppl. 1995;2:74–80.
93. Jung O, Schreiber JG, Geiger H, Pedrazzini T, Busse R, Brandes RP. gp91phox-containing NADPH oxidase mediates endothelial dysfunction in renovascular hypertension. Circulation. 2004;109:1795–801.
94. Landmesser U, Cai H, Dikalov S, McCann L, Hwang J, Jo H, et al. Role of p47(phox) in vascular oxidative stress and hypertension caused by angiotensin II. Hypertension. 2002;40: 511–5.
95. Droge W. Free radicals in the physiological control of cell function. Physiol Rev. 2002;82: 47–95.
96. Zinkevich NS, Gutterman DD. ROS-induced ROS release in vascular biology: redox-redox signaling. Am J Physiol Heart Circ Physiol. 2011;301:H647–53.
97. Su Q, Qin DN, Wang FX, Ren J, Li HB, Zhang M, et al. Inhibition of reactive oxygen species in hypothalamic paraventricular nucleus attenuates the renin-angiotensin system and proinflammatory cytokines in hypertension. Toxicol Appl Pharmacol. 2014;276:115–20.
98. Zimmerman MC, Lazartigues E, Lang JA, Sinnayah P, Ahmad IM, Spitz DR, et al. Superoxide mediates the actions of angiotensin II in the central nervous system. Circ Res. 2002;91:1038–45.
99. Cui W, Matsuno K, Iwata K, Ibi M, Katsuyama M, Kakehi T, et al. NADPH oxidase isoforms and anti-hypertensive effects of atorvastatin demonstrated in two animal models. J Pharmacol Sci. 2009;111:260–8.
100. Touyz RM. Reactive oxygen species, vascular oxidative stress, and redox signaling in hypertension: what is the clinical significance? Hypertension. 2004;44:248–52.
101. Costa CA, Amaral TA, Carvalho LC, Ognibene DT, da Silva AF, Moss MB, et al. Antioxidant treatment with tempol and apocynin prevents endothelial dysfunction and development of renovascular hypertension. Am J Hypertens. 2009;22:1242–9.
102. Schnackenberg CG, Wilcox CS. Two-week administration of tempol attenuates both hypertension and renal excretion of 8-Iso prostaglandin f2alpha. Hypertension. 1999;33:424–8.
103. Schnackenberg CG, Welch WJ, Wilcox CS. Normalization of blood pressure and renal vascular resistance in SHR with a membrane-permeable superoxide dismutase mimetic: role of nitric oxide. Hypertension. 1998;32:59–64.
104. Elks CM, Reed SD, Mariappan N, Shukitt-Hale B, Joseph JA, Ingram DK, et al. A blueberry-enriched diet attenuates nephropathy in a rat model of hypertension via reduction in oxidative stress. PLoS One. 2011;6, e24028.
105. Salonen RM, Nyyssonen K, Kaikkonen J, Porkkala-Sarataho E, Voutilainen S, Rissanen TH, et al. Six-year effect of combined vitamin C and E supplementation on atherosclerotic progression: the Antioxidant Supplementation in Atherosclerosis Prevention (ASAP) Study. Circulation. 2003;107:947–53.
106. Yusuf S, Dagenais G, Pogue J, Bosch J, Sleight P. Vitamin E supplementation and cardiovascular events in high-risk patients. The Heart Outcomes Prevention Evaluation Study Investigators. N Engl J Med. 2000;342:154–60.
107. Stephens NG, Parsons A, Schofield PM, Kelly F, Cheeseman K, Mitchinson MJ. Randomised controlled trial of vitamin E in patients with coronary disease: Cambridge Heart Antioxidant Study (CHAOS). Lancet. 1996;347:781–6.

108. Wang X, Skelley L, Wang B, Mejia A, Sapozhnikov V, Sun Z. AAV-based RNAi silencing of NADPH oxidase gp91(phox) attenuates cold-induced cardiovascular dysfunction. Hum Gene Ther. 2012;23:1016–26.
109. Eirin A, Ebrahimi B, Zhang X, Zhu XY, Woollard JR, He Q, et al. Mitochondrial protection restores renal function in swine atherosclerotic renovascular disease. Cardiovasc Res. 2014;103:461–72.
110. Eirin A, Li Z, Zhang X, Krier JD, Woollard JR, Zhu XY, et al. A mitochondrial permeability transition pore inhibitor improves renal outcomes after revascularization in experimental atherosclerotic renal artery stenosis. Hypertension. 2012;60:1242–9.
111. Nickenig G, Ostergren J, Struijker-Boudier H. Clinical evidence for the cardiovascular benefits of angiotensin receptor blockers. J Renin Angiotensin Aldosterone Syst. 2006;7 Suppl 1:S1–7.
112. Stafylas PC, Sarafidis PA, Grekas DM, Lasaridis AN. A cost-effectiveness analysis of angiotensin-converting enzyme inhibitors and angiotensin receptor blockers in diabetic nephropathy. J Clin Hypertens. 2007;9:751–9.
113. Tullis MJ, Caps MT, Zierler RE, Bergelin RO, Polissar N, Cantwell-Gab K, et al. Blood pressure, antihypertensive medication, and atherosclerotic renal artery stenosis. Am J Kidney Dis. 1999;33:675–81.
114. Losito A, Gaburri M, Errico R, Parente B, Cao PG. Survival of patients with renovascular disease and ACE inhibition. Clin Nephrol. 1999;52:339–43.
115. Losito A, Errico R, Santirosi P, Lupattelli T, Scalera GB, Lupattelli L. Long-term follow-up of atherosclerotic renovascular disease. Beneficial effect of ACE inhibition. Nephrol Dial Transplant. 2005;20:1604–9.
116. Hackam DG, Duong-Hua ML, Mamdani M, Li P, Tobe SW, Spence JD, et al. Angiotensin inhibition in renovascular disease: a population-based cohort study. Am Heart J. 2008;156: 549–55.
117. Chrysochou C, Foley RN, Young JF, Khavandi K, Cheung CM, Kalra PA. Dispelling the myth: the use of renin-angiotensin blockade in atheromatous renovascular disease. Nephrol Dial Transplant. 2012;27:1403–9.
118. Onuigbo MA, Onuigbo NT. Worsening renal failure in older chronic kidney disease patients with renal artery stenosis concurrently on renin angiotensin aldosterone system blockade: a prospective 50-month Mayo-Health-System clinic analysis. QJM. 2008;101:519–27.
119. Dincer Y, Sekercioglu N, Pekpak M, Gunes KN, Akcay T. Assessment of DNA oxidation and antioxidant activity in hypertensive patients with chronic kidney disease. Ren Fail. 2008;30: 1006–11.
120. Davies MG, Saad WE, Bismuth JX, Naoum JJ, Peden EK, Lumsden AB. Endovascular revascularization of renal artery stenosis in the solitary functioning kidney. J Vasc Surg. 2009;49:953–60.
121. Ziakka S, Ursu M, Poulikakos D, Papadopoulos C, Karakasis F, Kaperonis N, et al. Predictive factors and therapeutic approach of renovascular disease: four years' follow-up. Ren Fail. 2008;30:965–70.
122. Cooper CJ, Murphy TP, Cutlip DE, Jamerson K, Henrich W, Reid DM, et al. Stenting and medical therapy for atherosclerotic renal-artery stenosis. N Engl J Med. 2014;370:13–22.
123. Eirin A, Zhu XY, Urbieta-Caceres VH, Grande JP, Lerman A, Textor SC, et al. Persistent kidney dysfunction in swine renal artery stenosis correlates with outer cortical microvascular remodeling. Ren Physiol. 2011;300:F1394–401.
124. Eirin A, Ebrahimi B, Zhang X, Zhu XY, Tang H, Crane JA, et al. Changes in glomerular filtration rate after renal revascularization correlate with microvascular hemodynamics and inflammation in Swine renal artery stenosis. Circ Cardiovasc Interv. 2012;5:720–8.
125. Saad A, Herrmann SM, Crane J, Glockner JF, McKusick MA, Misra S, et al. Stent revascularization restores cortical blood flow and reverses tissue hypoxia in atherosclerotic renal artery stenosis but fails to reverse inflammatory pathways or glomerular filtration rate. Circ Cardiovasc Interv. 2013;6:428–35.

126. Wheatley K, Ives N, Gray R, Kalra PA, Moss JG, Baigent C, et al. Revascularization versus medical therapy for renal-artery stenosis. N Engl J Med. 2009;361:1953–62.
127. Eirin A, Zhu XY, Krier JD, Tang H, Jordan KL, Grande JP, et al. Adipose tissue-derived mesenchymal stem cells improve revascularization outcomes to restore renal function in swine atherosclerotic renal artery stenosis. Stem Cells. 2012;30:1030–41.
128. Zhuo W, Liao L, Xu T, Wu W, Yang S, Tan J. Mesenchymal stem cells ameliorate ischemia-reperfusion-induced renal dysfunction by improving the antioxidant/oxidant balance in the ischemic kidney. Urol Int. 2011;86:191–6.
129. Chen YT, Sun CK, Lin YC, Chang LT, Chen YL, Tsai TH, et al. Adipose-derived mesenchymal stem cell protects kidneys against ischemia-reperfusion injury through suppressing oxidative stress and inflammatory reaction. J Transl Med. 2011;9:51.
130. Hagiwara M, Shen B, Chao L, Chao J. Kallikrein-modified mesenchymal stem cell implantation provides enhanced protection against acute ischemic kidney injury by inhibiting apoptosis and inflammation. Hum Gene Ther. 2008;19:807–19.

Chapter 3
Oxidative Stress and Vascular Injury

Akshaar Brahmbhatt and Sanjay Misra

Introduction

Oxidative stress is responsible for aggravating vascular injury associated with atherosclerosis, chronic kidney disease (CKD), and end-stage renal disease (ESRD). The present chapter reviews the mechanisms responsible for oxidative stress contributing to vascular injury. We will discuss the role in hemodialysis vascular access failure, chronic kidney disease, and atherosclerosis.

Hemodialysis Vascular Access Failure

Vascular dialysis injury is characterized by the hyper-proliferation of cells leading to venous neointimal hyperplasia. Venous neointimal hyperplasia results from a process induced by mechanical forces, surgical trauma, complex hemodynamics, and repeat needle stick injury. These ultimately lead to vessel wall injury, which coupled with inflammation result in proliferation of fibroblasts, myofibroblasts, endothelial cells, and leukocytes. However, in the setting of end-stage renal disease, these cellular responses are much more pronounced due to the increased oxidative stress and the lack of antioxidant molecules and mechanisms present in ESRD.

Chronic kidney disease also causes lipid peroxidation, creates oxygen radicals, activates leukocytes, and induces cellular dysfunction through nucleic acid damage

A. Brahmbhatt, M.D. • S. Misra, M.D., F.S.I.R. F.A.H.A. (✉)
Vascular and Interventional Radiology Translational Laboratory, Mayo Clinic, Rochester, MN, USA

Division of Vascular and Interventional Radiology, Department of Radiology, Mayo Clinic, 200 First Street SW, Rochester, MN 55905, USA
e-mail: misra.sanjay@mayo.edu

M. Rodriguez-Porcel et al. (eds.), *Studies on Atherosclerosis*, Oxidative Stress in Applied Basic Research and Clinical Practice,
DOI 10.1007/978-1-4899-7693-2_3

[1–3]. As a whole, these stressors leave the vessel wall in a vulnerable state as evidence by the increased risk of atherosclerotic disease and venous neointimal hyperplasia (VNH) in the ESRD population. Thus, oxidative stress serves a synergistic role contributing to the molecular pathways that characterize vascular dialysis injury [3–8].

The surgical creation and repeat needle sticks cause local hypoxic trauma to the vessel wall. Studies have found variably increased oxidative stress among patients who are pre dialysis, on hemodialysis (HD), or on peritoneal dialysis (PD), suggesting that dialysis serves as an independent contributor of oxidative stress [9, 10]. Consistent with this, reactive oxygen species (ROS) can cause mitochondrial dysfunction. This further adds to the uremic environment by limiting oxidative respiration and thus causing more ROS [11, 12]. This process is evidenced by an increase in oxidants such as thiobarbituric acid reactive substances (TBARS), protein carbonyl content (PCO), 8-Hydroxy-2′-deoxyguanosine [HNE]), and lipid peroxidation (4-Hydroxy-2-Nonenal [8OHdG]), among others [13].

This is further worsened by the lack of antioxidants present in ESRD patients. Often increases in oxidative stress are counterbalanced by anti-oxidants and anti-oxidant generating mechanism, but this is not the case in ESRD patients. There have been many studies documenting the decreased levels of antioxidants such as superoxide dismutase, vitamin C, vitamin E, plasma sulfhydryl (P-SH), glutathione peroxidase, and systemic thiols in CKD patients [14]. Additionally there have been reductions in mediators of anti-oxidant pathways such as nuclear factor erythroid-2 [NF-E2]-related factor 2 (NrF2) [15]. These along with other factors have been linked to reduced antioxidant function in CKD patients [16]. All of these factors lead to cellular dysfunction through cellular damage and upregulation of inflammatory and proliferative pathways. This unbalanced level of increased oxidative stress among HD patients leads to an upregulation of cytokines which have also been implicated in atherosclerosis [3, 8, 17].

Chronic Kidney Disease

Oxidative stress and inflammation have been shown to have a direct relationship with worsening CKD. C reactive protein (CRP), interleukin 6 (Il-6), monocyte chemoattractant protein-1 (MCP-1), and several downstream pathways including activate protein kinase (AMPK) and B-cell lymphoma 2 (Bcl-2) correlate with different markers of oxidative stress [2, 18]. Additionally substrates of oxidative damage such as advanced glycosylation products (AGEs) can activate inflammatory receptors like RAGE and activate Il-6, CRP, and monocytes [3].

These changes prime the body's inflammatory mechanisms and enhances local inflammation caused by macrophages and infiltrating lymphocytes at the hemodialysis access site secondary to surgical and needle trauma [7]. One major cytokine is tumor necrosis factor-α (TNF-α), which acts through a nuclear factor-kappa-B

(NF-κB) pathway via several receptors, the receptor for advanced glycation end products (RAGE), and TNF-receptor one (TNF-R1) [19–21]. In fact, RAGE has been found to be up regulated in other models of vascular damage and have been linked to vascular cell adhesion protein 1 (VCAM-1), a marker implicated in ECM deposition [22]. TNF-α has also been found to induce Interleukin 1 (IL-1), prostaglandin E2 (PGE2), and TNF-α from smooth muscle cells (SMCs) and endothelial cells (ECs), where it serves to propagate local inflammatory and proliferative changes [23] and stimulates proliferative changes in fibroblasts [24]. Furthermore, specific polymorphisms in TNF-α have been linked to differing risk for AVF thrombosis [25]. It is likely that this is modulated in some part by ROS activating concurrent NF-κB pathways [1, 26].

In response to oxidative stress, venous endothelial cells have been shown to release endogenous damage-associated molecular patterns, which act through toll-like receptors (TLR), further propagating the immune response through NF-κB. In addition to TLR there are inflammasomes, multi-protein complexes with sensor and adaptor proteins that can modulate inflammatory conditions via regulation of pathways such as NF-κB and p38 but can also directly active cytokines. One prominent example of this is the NACHT, LRR and PYD domains-containing protein 3 (NALP3) and apoptosis-associated speck-like protein containing CARD (ASC) which activates caspase-1 [27–29].

The local macrophage response is thought to be mediated by migration inhibitory factor (MIF) [30, 31], which drives inflammatory cells towards the intimal cell layer. This migration is accompanied by activation of CD74, chemokine (C-X-C motif) receptor 2 (CXCR2), and chemokine (C-X-C motif) receptor 4 (CXCR4) receptors [31]. These receptors act through an extracellular signal-regulated kinases (ERK) and p38 pathways to up regulate other pro inflammatory cytokines interleukin 8 (IL-8), monocyte chemotactic protein-1 (MCP-1). Both Il-8 and MCP1/CCL2 induce the activation and migration of monocytes, memory T lymphocytes, and natural killer cells (NK) to the site of vascular injury through chemokine (C-X-C motif) receptors [32]. These local and systemic inflammatory shifts also activate synergistic proliferative cytokines such as transforming growth factor beta (TGF-β) and insulin like growth factor-1 (IGF-1) [33].

Beyond the inflammatory changes caused by uremia, there are also many direct effects of reactive oxygen species (ROS), reactive nitrogen species (RNS), and hypoxia. These pathways are mediated by the dysregulation of strong cell signaling molecules including hypoxia inducible factor 1α (HIF-1α), vascular endothelial growth factors (VEGFs), TGF-β1, heme oxygenase-1 (HO-1), and -2 (HO-2) [34–37].

HIF-1α and its downstream mediators have been found to be increased in both animal and human examples of VNH and venous hypertension [34, 38]. HIF-1α expression is involved in many molecular mechanisms, which lead to increased inflammation, cell mitosis, and angiogenesis [39]. The major downstream cytokines linked to HIF-1α are of the VEGF family, which are thought to cause the arterialization and negative remodeling of the vessel wall.

VEGF-A acts through VEGF-R1 and VEGF-R2, which are receptors that act via tyrosine kinase pathways promoting endothelial growth. It is important to note that VEGF-R2 is more commonly expressed on vascular tissue [40, 41], where it can induce SMC proliferation via ERK and AKT pathways [42]. VEGF-R1 is also implicated in the activation of macrophages [43]. VEGF-A has also been shown to modulate the ECM via the activation of matrixmetalloprotease-9 (MMP-9), and matrix metalloproteinase-2 (MMP-2). ADAMTS-1 is another matrix metalloproteinase up regulated in models of AVF failure [34]. The ECM changes work in concert with the pro-mitotic effects of VEGF-A [39].

In contrast to VEGF-A, heme oxygenase-1 and -2 serve in protective roles against negative vascular remodeling. HO-1 is inducible where as HO-2 is already present in vessel. Both appear to reduce MMP-9, while only HO-1 reduces MMP-2. In addition, HO-1 has been shown to reduce the effects of plasminogen activator inhibitor-1 and MCP-1. Both likely serve to protect cells against apoptosis and are anti-inflammatory [44].

There are many isoforms of VEGF and these unique forms allow for modulation of angiogenesis and proliferation [45]. The VEGF-A165b isoform is anti-angiogenic and its splice variant VEGF-A165a is pro-angiogenic. Both have been studied in models of peripheral artery disease [46]. The inflammatory milieu created by TNF-α leads to more proximal pro-angiogenic VEGF isoforms. In contrast, TGF-β1 leads to more distal splice site variants, likely via p38 MAPK-Clk/sty kinases [47]. In addition, previous work has shown that VEGF and TGF-β1 likely influence each other through a SMAD3 pathway. Upregulation of TGF-β/Smad3 has been shown to decrease apoptosis and increase secretion of VEGF-A in SMCs [48–50]. In addition to splice variants, inherent polymorphisms the such a VEGF-936C/C genotype have been linked to increased risk for AVF failure [51].

TGF-β1 has been shown to enhance proliferation, fibrosis, and thrombosis. In this setting TGF-β acts via a Smad3 mechanism. It has been shown to decrease apoptosis and increase secretion of VEGF-A in SMCs. TGF-β has also been shown to increase ECM deposition [33]. Downstream core fucosyltransferases linked to TGF-β have also been shown to be involved in cancer models where it modulates cellular adhesion, mitosis, and apoptosis [52, 53]. The proliferative nature of TGF-β is dependent on the local microenvironment as well as differences in genetic polymorphisms [36, 48–50]. Platelet-derived growth factor is another cytokine induced by hypoxia. It serves in a proliferative role by inducing proliferation in myofibroblasts via AKT and ERK pathways [42, 54, 55].

In addition to the inflammatory and proliferative pathways influenced by oxidative stress, its effects on lipid peroxidation, local fibroblasts, and other cells are also significant in hemodialysis access failure. These changes are implicated in access failure and in atherosclerosis. There have been several studies that demonstrate increased intima and media thickness arterially and dyslipidemia in ESRD patients [56]. However, studies have shown that increased IMT may precede CKD [57]. The data suggests that those with more severe CKD are at higher risk for cardiovascular morbidity and mortality. However, patients with CKD often have several other risk factors confounding these studies [58, 59].

Atherosclerosis

The complications resulting from atherosclerosis are responsible for the majority of cardiovascular problems in the world. These are manifested by the formation of atherosclerotic plaque. An early step in plaque formation is the production of oxidized low-density lipoprotein (Ox-LDL) and its effect on the endothelium. Ox-LDLs can damage endothelial cells along with smooth muscle cells. This cascade leads to the deposition of monocytes and lymphocytes on the endothelium. This is mediated in part by the induction of several inflammatory cytokines including increased expression of adhesion molecules such as P-selectin and chemotactic factors such as MCP-1 and macrophage colony stimulating factor (CSF). Oxidative damage also propagates atherosclerosis by impairing HDL-mediated reverse cholesterol transport [60]. A combination of these two factors are in large part responsible for the increased atherosclerosis and cardiovascular risk seen in CKD patients [61].

Oxidative stress can occur secondary to uremia and hypoxia but also by dysregulation of nicotinamide adenine dinucleotide (phosphate) oxidases (NOX). There are several NOXs present in the vasculature and elsewhere throughout the body. They are named by their intramembranous anchoring portions. Nox1 expression has been seen in vascular smooth muscle cells and is thought to play a regulatory role in blood pressure control in relation to angiotensin II. However, there is incomplete evidence as to whether NOX1 is expressed constituently or only in pathologic conditions. However, it does appear that NOX1 is expressed in VSMCS after injury and contributes to oxidative stress by producing O_2^- [62–64]. There are several other NOX types 2,3,4,5, Duox1, and 2. Out of these, NOX4 related to the vasculature, because it is expressed on vascular endothelial cells in the nucleus. It is suspected to possibly play a role in counterbalancing NO production and activity. It is also present on fibroblasts and has been shown aid in TGF-B1-driven transformation of these cells into myofibroblasts [65, 66].

While several other NOX isoforms have been shown to be expressed in vascular tissue in a variety of pathological contexts (including atherosclerosis), most have not been extensively investigated in the context of atherosclerosis accelerated by CKD or ESRD. In addition, there several enzymes and micro RNAs, which modulate NOX expression and activity, present on several cell types related to atherosclerosis. The nature of these concerning atherosclerosis has yet to be fully elucidated [63, 67, 68].

Interestingly, oxidative stress itself can result in the increased production of NADPH, NOX OX, and xanthine oxidases (XO). In experiments performed in cell culture, NOX expression can be increased by several inflammatory cytokines including TNF-α, angiotensin II (Ang II), thrombin, and platelet-derived growth factor (PDGF) [69]. Mice deficient for p47Phox a enzyme needed for NAD(P)H oxidase were found to have less superoxide and reduced atherosclerosis compared to wild types. There is evidence for NAD(P)H oxidase in human atherosclerosis specimens. In vitro, high levels of agents which induce NAD(P)H oxidase have been observed including Ang II. This has been seen at the shoulder regions of atherosclerotic plaques where plaque rupture can occur leading to acute myocardial infarction [69].

The oxidative stressors produced by enzymes such as superoxide anion can inactivate beneficial anti-atherosclerotic radicals like NO. Additionally they cause the formation of lysophosphatidylcholine and other reacted molecules that can induce atherosclerosis [63]. Hydrogen peroxide, a more stable ROS, can act on smooth muscle cells inducing mitosis and migration [70].

These stressors can also activate other enzymes such as lipoxygenases (LOs). Of these LO-5 is the best studied in vasculature and leads to leukotriene production and increases inflammation by promoting chemotaxis and vascular permeability [71]. Genetic association studies have tied the activator for LO-5 and Leukotriene A4 hydrolase to cardiovascular disease, suggesting a present but possibly limited role for leukotrienes in atherosclerosis. There have been mixed results with studies looking at reducing atherosclerosis via LO inhibition [72–74].

Kynurenine (KYN) and its metabolites kynurenic acid (KYNA) and quinolinic acid (QA) are three markers of oxidative stress that have been shown to be increased in uremia and atherosclerosis. KYN is a byproduct of tryptophan catabolization by indoleamine 2,3-dioxygenase. Increased levels KYN and QA have been associated with increased inflammation, intima-media thickness (IMT) values, anemia, and other markers of reactive oxygen species. Increase activity of indoleamine 2,3-dioxygenase as measured by increased ratios of KYN/tryptophan have been correlated with increased time on hemodialysis, total lymphocytes, CRP, ankle-brachial pressure, and carotid artery intima-media thickness in HD patients. While work has linked the kynurenine pathway to oxidative stress and atherosclerotic changes in the setting of CKD, the mechanisms are still unknown. It is likely that they relate to immune modulators such as Interferon-γ and IL-6 [75].

Several other factors tie together atherosclerosis and oxidative vascular damage. For example, obesity has been linked to increased atherosclerosis, elevated CRP but also increased risk of primary AVF failure [76]. Other cellular phenotypes, inflammatory cells, collagen expression have also been linked to ESRD, atherosclerosis, and vascular dialysis injury [77, 78]. Overall, it makes sense that the factors involved in vascular access injury are the same as those for atherosclerosis and ESRD as both compromise the homeostasis of cells needed to maintain healthy vasculature.

There have been many studies looking at the effects of therapies to reduce inflammation and oxidative burden. Statins have shown some promise, but results are mixed [79, 80]. Studies looking at vitamin E, *N*-acetylcysteine (NAC), and caffeic acid have had limited efficacy [2]. Several other studies have shown mixed results. For example, pentoxifylline has shown to improve anemia but does not affect oxidative metabolism [81]. A few studies have demonstrated a reduction in ROS, but did not result in clinically significant changes [82]. In addition to macrophages and local fibroblasts, it is likely that mesenchymal-derived cells in bone are also affected by uremia and oxidative stress. It is unclear if mesenchymal cells in the bone play a role in vascular injury [83, 84].

Conclusion

The deleterious effects of oxidative stress extend far beyond hemodialysis access function. They are found to be involved in chronic kidney disease and atherosclerosis. Therapies aimed at reducing these effects could result in drastic improvements in the mortality and morbidity associated with increased oxidative stress across a broad range of human disease conditions.

Acknowledgments This work was funded by a HL098967 (SM) from the National Heart, Lung, And Blood Institute.

References

1. Tucker PS, Scanlan AT, Dalbo VJ. Chronic Kidney Disease Influences Multiple Systems: Describing the Relationship between Oxidative Stress, Inflammation, Kidney Damage, and Concomitant Disease. Oxid Med Cell Longev. 2015;2015:806358. doi:10.1155/2015/806358.
2. Oberg BP, McMenamin E, Lucas FL, McMonagle E, Morrow J, Ikizler TA, et al. Increased prevalence of oxidant stress and inflammation in patients with moderate to severe chronic kidney disease. Kidney Int. 2004;65(3):1009–16. doi:10.1111/j.1523-1755.2004.00465.x.
3. Himmelfarb J, Stenvinkel P, Ikizler TA, Hakim RM. The elephant in uremia: oxidant stress as a unifying concept of cardiovascular disease in uremia. Kidney Int. 2002;62(5):1524–38. doi:10.1046/j.1523-1755.2002.00600.x.
4. Kokubo T, Ishikawa N, Uchida H, Chasnoff SE, Xie X, Mathew S, et al. CKD accelerates development of neointimal hyperplasia in arteriovenous fistulas. J Am Soc Nephrol. 2009;20(6):1236–45. doi:10.1681/asn.2007121312.
5. Yang B, Vohra PK, Janardhanan R, Misra KD, Misra S. Expression of profibrotic genes in a murine remnant kidney model. J Vasc Interv Radiol. 2011;22 12, 1765–1772.e1761. doi:10.1016/j.jvir.2011.08.026
6. Lee T, Chauhan V, Krishnamoorthy M, Wang Y, Arend L, Mistry MJ, et al. Severe venous neointimal hyperplasia prior to dialysis access surgery. Nephrol Dial Transplant. 2011;26(7):2264–70. doi:10.1093/ndt/gfq733.
7. Liang A, Wang Y, Han G, Truong L, Cheng J. Chronic kidney disease accelerates endothelial barrier dysfunction in a mouse model of an arteriovenous fistula. Am J Physiol Renal Physiol. 2013;304(12):F1413–20. doi:10.1152/ajprenal.00585.2012.
8. Cachofeiro V, Goicochea M, de Vinuesa SG, Oubina P, Lahera V, Luno J. Oxidative stress and inflammation, a link between chronic kidney disease and cardiovascular disease. Kidney Int Suppl. 2008;111:S4–9. doi:10.1038/ki.2008.516.
9. Puchades MJ, Saez G, Munoz MC, Gonzalez M, Torregrosa I, Juan I, et al. Study of oxidative stress in patients with advanced renal disease and undergoing either hemodialysis or peritoneal dialysis. Clin Nephrol. 2013;80(3):177–86. doi:10.5414/cn107639.
10. Ansarihadipour H, Dorostkar H. Comparison of plasma oxidative biomarkers and conformational modifications of hemoglobin in patients with diabetes on hemodialysis. Iran Red Crescent Med J. 2014;16(11), e22045. doi:10.5812/ircmj.22045.
11. Tschopp J. Mitochondria: Sovereign of inflammation? Eur J Immunol. 2011;41(5):1196–202. doi:10.1002/eji.201141436.
12. Yazdi PG, Moradi H, Yang JY, Wang PH, Vaziri ND. Skeletal muscle mitochondrial depletion and dysfunction in chronic kidney disease. Int J Clin Exp Med. 2013;6(7):532–9.
13. Wasse H, Huang R, Naqvi N, Smith E, Wang D, Husain A. Inflammation, oxidation and venous neointimal hyperplasia precede vascular injury from AVF creation in CKD patients. J Vasc Access. 2012;13(2):168–74. doi:10.5301/jva.5000024.

14. Dursun B, Dursun E, Suleymanlar G, Ozben B, Capraz I, Apaydin A, et al. Carotid artery intima-media thickness correlates with oxidative stress in chronic haemodialysis patients with accelerated atherosclerosis. Nephrol Dial Transplant. 2008;23(5):1697–703. doi:10.1093/ndt/gfm906.
15. Ruiz S, Pergola PE, Zager RA, Vaziri ND. Targeting the Transcription Factor Nrf2 to Ameliorate Oxidative Stress and Inflammation in Chronic Kidney Disease. Kidney Int. 2013;83(6):1029–41. doi:10.1038/ki.2012.439.
16. Locatelli F, Canaud B, Eckardt KU, Stenvinkel P, Wanner C, Zoccali C. Oxidative stress in end-stage renal disease: an emerging threat to patient outcome. Nephrol Dial Transplant. 2003;18(7):1272–80.
17. Granata S, Zaza G, Simone S, Villani G, Latorre D, Pontrelli P, et al. Mitochondrial dysregulation and oxidative stress in patients with chronic kidney disease. BMC Genomics. 2009;10:388. doi:10.1186/1471-2164-10-388.
18. Lee J, Giordano S, Zhang J. Autophagy, mitochondria and oxidative stress: cross-talk and redox signalling. Biochem J. 2012;441(2):523–40. doi:10.1042/bj20111451.
19. Feldman HI, Joffe M, Rosas SE, Burns JE, Knauss J, Brayman K. Predictors of successful arteriovenous fistula maturation. Am J Kidney Dis. 2003;42(5):1000–12.
20. Huijbregts HJT, Bots ML, Wittens CHA, Schrama YC, Moll FL, Blankestijn PJ, et al. Hemodialysis arteriovenous fistula patency revisited: results of a prospective, multicenter initiative. Clin J Am Soc Nephrol. 2008;3(3):714–9. doi:10.2215/CJN.02950707.
21. Gupta S, Gambhir JK, Kalra O, Gautam A, Shukla K, Mehndiratta M, et al. Association of biomarkers of inflammation and oxidative stress with the risk of chronic kidney disease in Type 2 diabetes mellitus in North Indian population. J Diabetes Complications. 2013;27(6):548–52. doi:10.1016/j.jdiacomp.2013.07.005.
22. Takeda R, Suzuki E, Satonaka H, Oba S, Nishimatsu H, Omata M, et al. Blockade of endogenous cytokines mitigates neointimal formation in obese Zucker rats. Circulation. 2005;111(11):1398–406. doi:10.1161/01.cir.0000158482.83179.db.
23. Vassalotti JA, Jennings WC, Beathard GA, Neumann M, Caponi S, Fox CH, et al. Fistula first breakthrough initiative: targeting catheter last in fistula first. Semin Dial. 2012;25(3):303–10. doi:10.1111/j.1525-139X.2012.01069.x.
24. Dixon BS. Why don't fistulas mature? Kidney Int. 2006;70(8):1413–22. doi:10.1038/sj.ki.5001747.
25. Sener EF, Taheri S, Korkmaz K, Zararsiz G, Serhatlioglu F, Unal A, et al. Association of TNF-alpha −308 G > A and ACE I/D gene polymorphisms in hemodialysis patients with arteriovenous fistula thrombosis. Int Urol Nephrol. 2014;46(7):1419–25. doi:10.1007/s11255-013-0580-2.
26. Guijarro C, Egido J. Transcription factor-kappa B (NF-kappa B) and renal disease. Kidney Int. 2001;59(2):415–24. doi:10.1046/j.1523-1755.2001.059002415.x.
27. Carbo C, Arderiu G, Escolar G, Fuste B, Cases A, Carrascal M, et al. Differential expression of proteins from cultured endothelial cells exposed to uremic versus normal serum. Am J Kidney Dis. 2008;51(4):603–12. doi:10.1053/j.ajkd.2007.11.029.
28. Martin-Rodriguez S, Caballo C, Gutierrez G, Vera M, Cruzado JM, Cases A, et al. TLR4 and NALP3 inflammasome in the development of endothelial dysfunction in uraemia. Eur J Clin Invest. 2015;45(2):160–9. doi:10.1111/eci.12392.
29. Schroder K, Tschopp J. The inflammasomes. Cell. 2010;140(6):821–32. doi:10.1016/j.cell.2010.01.040.
30. Misra S, Fu AA, Rajan DK, Juncos LA, McKusick MA, Bjarnason H, et al. Expression of hypoxia inducible factor-1 alpha, macrophage migration inhibition factor, matrix metalloproteinase-2 and −9, and their inhibitors in hemodialysis grafts and arteriovenous fistulas. J Vasc Interv Radiol. 2008;19(2 Pt 1):252–9. doi:10.1016/j.jvir.2007.10.031.
31. Asare Y, Schmitt M, Bernhagen J. The vascular biology of macrophage migration inhibitory factor (MIF). Expression and effects in inflammation, atherogenesis and angiogenesis. Thromb Haemost. 2013;109(3):391–8. doi:10.1160/th12-11-0831.
32. Deshmane SL, Kremlev S, Amini S, Sawaya BE. Monocyte chemoattractant protein-1 (MCP-1): an overview. J Interferon Cytokine Res. 2009;29(6):313–26. doi:10.1089/jir.2008.0027.

33. Stracke S, Konner K, Köstlin I, Friedl R, Jehle PM, Hombach V, et al. Increased expression of TGF-beta1 and IGF-I in inflammatory stenotic lesions of hemodialysis fistulas. Kidney Int. 2002;61(3):1011–9. doi:10.1046/j.1523-1755.2002.00191.x.
34. Misra S, Shergill U, Yang B, Janardhanan R, Misra KD. Increased expression of HIF-1alpha, VEGF-A and its receptors, MMP-2, TIMP-1, and ADAMTS-1 at the venous stenosis of arteriovenous fistula in a mouse model with renal insufficiency. J Vasc Interv Radiol. 2010;21(8):1255–61. doi:10.1016/j.jvir.2010.02.043.
35. Misra S, Fu AA, Puggioni A, Glockner JF, Rajan DK, McKusick MA, et al. Increased expression of hypoxia-inducible factor-1 alpha in venous stenosis of arteriovenous polytetrafluoroethylene grafts in a chronic renal insufficiency porcine model. J Vasc Interv Radiol. 2008;19(2 Pt 1):260–5. doi:10.1016/j.jvir.2007.10.029.
36. Heine GH, Ulrich C, Sester U, Sester M, Kohler H, Girndt M. Transforming growth factor beta1 genotype polymorphisms determine AV fistula patency in hemodialysis patients. Kidney Int. 2003;64(3):1101–7. doi:10.1046/j.1523-1755.2003.00176.x.
37. Misra S, Fu AA, Puggioni A, Karimi KM, Mandrekar JN, Glockner JF, et al. Increased shear stress with upregulation of VEGF-A and its receptors and MMP-2, MMP-9, and TIMP-1 in venous stenosis of hemodialysis grafts. Am J Physiol Heart Circ Physiol. 2008;294(5):H2219–2230. doi:10.1152/ajpheart.00650.2007.
38. Zhu Y, Lawton MT, Du R, Shwe Y, Chen Y, Shen F, et al. Expression of hypoxia-inducible factor-1 and vascular endothelial growth factor in response to venous hypertension. Neurosurgery. 2006;59(3):687–96. doi:10.1227/01.neu.0000228962.68204.cf. discussion 687–696.
39. Semenza GL. Targeting HIF-1 for cancer therapy. Nat Rev Cancer. 2003;3(10):721–32. doi:10.1038/nrc1187.
40. Shibuya M. Differential roles of vascular endothelial growth factor receptor-1 and receptor-2 in angiogenesis. J Biochem Mol Biol. 2006;39(5):469–78.
41. Huusko J, Merentie M, Dijkstra MH, Ryhänen MM, Karvinen H, Rissanen TT, et al. The effects of VEGF-R1 and VEGF-R2 ligands on angiogenic responses and left ventricular function in mice. Cardiovasc Res. 2010;86(1):122–30. doi:10.1093/cvr/cvp382.
42. Wan J, Lata C, Santilli A, Green D, Roy S, Santilli S. Supplemental Oxygen Reverses Hypoxia Induced Smooth Muscle Cell Proliferation by Modulating HIF-alpha and VEGF Levels in a Rabbit Arteriovenous Fistula Model. Ann Vasc Surg. 2014;28(3):725–36. doi:10.1016/j.avsg.2013.10.007.
43. Ohtani K, Egashira K, Hiasa K, Zhao Q, Kitamoto S, Ishibashi M, et al. Blockade of vascular endothelial growth factor suppresses experimental restenosis after intraluminal injury by inhibiting recruitment of monocyte lineage cells. Circulation. 2004;110(16):2444–52. doi:10.1161/01.cir.0000145123.85083.66.
44. Kang L, Grande JP, Farrugia G, Croatt AJ, Katusic ZS, Nath KA. Functioning of an arteriovenous fistula requires heme oxygenase-2. Am J Physiol Renal Physiol. 2013;305(4):F545–552. doi:10.1152/ajprenal.00234.2013.
45. Vempati P, Popel AS, Mac Gabhann F. Extracellular regulation of VEGF: isoforms, proteolysis, and vascular patterning. Cytokine Growth Factor Rev. 2014;25(1):1–19. doi:10.1016/j.cytogfr.2013.11.002.
46. Kikuchi R, Nakamura K, MacLauchlan S, Ngo DT, Shimizu I, Fuster JJ. An antiangiogenic isoform of VEGF-A contributes to impaired vascularization in peripheral artery disease. Nat Med. 2014;20(12):1464–71. doi:10.1038/nm.3703.
47. Nowak DG, Woolard J, Amin EM, Konopatskaya O, Saleem MA, Churchill AJ, et al. Expression of pro- and anti-angiogenic isoforms of VEGF is differentially regulated by splicing and growth factors. J Cell Sci. 2008;121(Pt 20):3487–95. doi:10.1242/jcs.016410.
48. Geng L, Chaudhuri A, Talmon G, Wisecarver JL, Wang J. TGF-Beta suppresses VEGFA-mediated angiogenesis in colon cancer metastasis. PLoS One. 2013;8(3), e59918. doi:10.1371/journal.pone.0059918.
49. Nakagawa T, Li JH, Garcia G, Mu W, Piek E, Böttinger EP, et al. TGF-beta induces proangiogenic and antiangiogenic factors via parallel but distinct Smad pathways. Kidney Int. 2004;66(2):605–13. doi:10.1111/j.1523-1755.2004.00780.x.

50. Shi X, Guo LW, Seedial SM, Si Y, Wang B, Takayama T, et al. TGF-β/Smad3 inhibit vascular smooth muscle cell apoptosis through an autocrine signaling mechanism involving VEGF-A. Cell Death Dis. 2014;5(7), e1317. doi:10.1038/cddis.2014.282.
51. Candan F, Yildiz G, Kayatas M. Role of the VEGF 936 gene polymorphism and VEGF-A levels in the late-term arteriovenous fistula thrombosis in patients undergoing hemodialysis. Int Urol Nephrol. 2014;46(9):1815–23. doi:10.1007/s11255-014-0711-4.
52. Zhao YY, Takahashi M, Gu JG, Miyoshi E, Matsumoto A, Kitazume S, et al. Functional roles of N-glycans in cell signaling and cell adhesion in cancer. Cancer Sci. 2008;99(7):1304–10. doi:10.1111/j.1349-7006.2008.00839.x.
53. Shen N, Lin H, Wu T, Wang D, Wang W, Xie H, et al. Inhibition of TGF-beta1-receptor post-translational core fucosylation attenuates rat renal interstitial fibrosis. Kidney Int. 2013;84(1):64–77. doi:10.1038/ki.2013.82.
54. Simone S, Loverre A, Cariello M, Divella C, Castellano G, Gesualdo L, et al. Arteriovenous fistula stenosis in hemodialysis patients is characterized by an increased adventitial fibrosis. J Nephrol. 2014;27(5):555–62. doi:10.1007/s40620-014-0050-7.
55. Lata C, Green D, Wan J, Roy S, Santilli SM. The role of short-term oxygen administration in the prevention of intimal hyperplasia. J Vasc Surg. 2013;58(2):452–9. doi:10.1016/j.jvs.2012.11.041.
56. Hinderliter A, Padilla RL, Gillespie BW, Levin NW, Kotanko P, Kiser M, et al. Association of carotid intima-media thickness with cardiovascular risk factors and patient outcomes in advanced chronic kidney disease: the RRI-CKD study. Clin Nephrol. 2015;84(7):10–20. doi:10.5414/cn108494.
57. Shimizu M, Furusyo N, Mitsumoto F, Takayama K, Ura K, Hiramine S, et al. Subclinical carotid atherosclerosis and triglycerides predict the incidence of chronic kidney disease in the Japanese general population: results from the Kyushu and Okinawa Population Study (KOPS). Atherosclerosis. 2015;238(2):207–12. doi:10.1016/j.atherosclerosis.2014.12.013.
58. Sarnak MJ, Levey AS, Schoolwerth AC, Coresh J, Culleton B, Hamm LL, et al. Kidney disease as a risk factor for development of cardiovascular disease: a statement from the American Heart Association Councils on Kidney in Cardiovascular Disease, High Blood Pressure Research, Clinical Cardiology, and Epidemiology and Prevention. Circulation. 2003;108(17):2154–69. doi:10.1161/01.cir.0000095676.90936.80.
59. Culleton BF, Hemmelgarn BR. Is chronic kidney disease a cardiovascular disease risk factor? Semin Dial. 2003;16(2):95–100.
60. Vaziri ND. Lipotoxicity and impaired high density lipoprotein-mediated reverse cholesterol transport in chronic kidney disease. J Ren Nutr. 2010;20(5 Suppl):S35–43. doi:10.1053/j.jrn.2010.05.010.
61. Navab KD, Elboudwarej O, Gharif M, Yu J, Hama SY, Safarpour S, et al. Chronic inflammatory disorders and accelerated atherosclerosis: chronic kidney disease. Curr Pharm Des. 2011;17(1):17–20.
62. Szöcs K, Lassègue B, Sorescu D, Hilenski LL, Valppu L, Couse TL, et al. Upregulation of Nox-based NAD(P)H oxidases in restenosis after carotid injury. Arterioscler Thromb Vasc Biol. 2002;22(1):21–7.
63. Yokoyama M, Inoue N, Kawashima S. Role of the vascular NADH/NADPH oxidase system in atherosclerosis. Ann N Y Acad Sci. 2000;902:241–7. discussion 247–248.
64. Jacobson GM, Dourron HM, Liu J, Carretero OA, Reddy DJ, Andrzejewski T, et al. Novel NAD(P)H oxidase inhibitor suppresses angioplasty-induced superoxide and neointimal hyperplasia of rat carotid artery. Circ Res. 2003;92(6):637–43. doi:10.1161/01.res.0000063423.94645.8a.
65. Bondi CD, Manickam N, Lee DY, Block K, Gorin Y, Abboud HE, et al. NAD(P)H oxidase mediates TGF-β1–induced activation of kidney myofibroblasts. J Am Soc Nephrol. 2010;21(1):93–102. doi:10.1681/ASN.2009020146.
66. Cucoranu I, Clempus R, Dikalova A, Phelan PJ, Ariyan S, Dikalov S, et al. NAD(P)H oxidase 4 mediates transforming growth factor-beta1-induced differentiation of cardiac fibroblasts into myofibroblasts. Circ Res. 2005;97(9):900–7. doi:10.1161/01.res.0000187457.24338.3d.

67. Rana I, Velkoska E, Patel SK, Burrell LM, Charchar FJ. MicroRNAs mediate the cardioprotective effect of angiotensin converting enzyme inhibition in acute kidney injury. Am J Physiol Renal Physiol. 2015;309(11):F943–54. doi:10.1152/ajprenal.00183.2015.
68. Geiszt M. NADPH oxidases: new kids on the block. Cardiovasc Res. 2006;71(2):289–99. doi:10.1016/j.cardiores.2006.05.004.
69. Pawlak K, Brzosko S, Mysliwiec M, Pawlak D. Kynurenine, quinolinic acid the new factors linked to carotid atherosclerosis in patients with end-stage renal disease. Atherosclerosis. 2009;204(2):561–6. doi:10.1016/j.atherosclerosis.2008.10.002.
70. Madamanchi NR, Vendrov A, Runge MS. Oxidative stress and vascular disease. Arterioscler Thromb Vasc Biol. 2005;25(1):29–38. doi:10.1161/01.atv.0000150649.39934.13.
71. Griendling KK, Sorescu D, Lassegue B, Ushio-Fukai M. Modulation of protein kinase activity and gene expression by reactive oxygen species and their role in vascular physiology and pathophysiology. Arterioscler Thromb Vasc Biol. 2000;20(10):2175–83.
72. De Caterina R, Zampolli A. From asthma to atherosclerosis-5-lipoxygenase, leukotrienes, and inflammation. N Engl J Med. 2004;350(1):4–7. doi:10.1056/NEJMp038190.
73. Crosslin DR, Shah SH, Nelson SC, Haynes CS, Connelly JJ, Gadson S, et al. Genetic effects in the leukotriene biosynthesis pathway and association with atherosclerosis. Hum Genet. 2009;125(2):217–29. doi:10.1007/s00439-008-0619-0.
74. Yang LX, Heng XH, Guo RW, Si YK, Qi F, Zhou XB. Atorvastatin inhibits the 5-lipoxygenase pathway and expression of CCL3 to alleviate atherosclerotic lesions in atherosclerotic ApoE knockout mice. J Cardiovasc Pharmacol. 2013;62(2):205–11. doi:10.1097/FJC.0b013e3182967fc0.
75. Back M. Inhibitors of the 5-lipoxygenase pathway in atherosclerosis. Curr Pharm Des. 2009;15(27):3116–32.
76. Kim JK, Jeong JH, Song YR, Kim HJ, Lee WY, Kim KI, et al. Obesity-related decrease in intraoperative blood flow is associated with maturation failure of radiocephalic arteriovenous fistula. J Vasc Surg. 2015. doi:10.1016/j.jvs.2015.05.008.
77. Bai Y, Zhang J, Xu J, Cui L, Zhang H, Zhang S. Alteration of type I collagen in the radial artery of patients with end-stage renal disease. Am J Med Sci. 2015;349(4):292–7. doi:10.1097/maj.0000000000000408.
78. Rekhter MD, Zhang K, Narayanan AS, Phan S, Schork MA, Gordon D. Type I collagen gene expression in human atherosclerosis. Localization to specific plaque regions. Am J Pathol. 1993;143(6):1634–48.
79. Fassett RG, Robertson IK, Ball MJ, Geraghty DP, Coombes JS. Effects of atorvastatin on oxidative stress in chronic kidney disease. Nephrology (Carlton). 2015. doi:10.1111/nep.12502.
80. Kadowaki D, Anraku M, Sakaya M, Hirata S, Maruyama T, Otagiri M. Olmesartan protects endothelial cells against oxidative stress-mediated cellular injury. Clin Exp Nephrol. 2015. doi:10.1007/s10157-015-1111-5.
81. Zhang L, Coombes J, Pascoe EM, Badve SV, Dalziel K, Cass A, et al. The effect of pentoxifylline on oxidative stress in chronic kidney disease patients with erythropoiesis-stimulating agent hyporesponsiveness: sub-study of the HERO trial. Redox Rep. 2015. doi:10.1179/1351000215y.0000000022.
82. DuPont JJ, Ramick MG, Farquhar WB, Townsend RR, Edwards DG. NADPH oxidase-derived reactive oxygen species contribute to impaired cutaneous microvascular function in chronic kidney disease. Am J Physiol Renal Physiol. 2014;306(12):F1499–506. doi:10.1152/ajprenal.00058.2014.
83. Hruska KA, Mathew S, Memon I, Saab G. The pathogenesis of vascular calcification in the chronic kidney disease mineral bone disorder (CKD-MBD): the links between bone and the vasculature. Semin Nephrol. 2009;29(2):156–65. doi:10.1016/j.semnephrol.2009.01.008.
84. Yamanouchi D, Takei Y, Komori K. Balanced mineralization in the arterial system: possible role of osteoclastogenesis/osteoblastogenesis in abdominal aortic aneurysm and stenotic disease. Circ J. 2012;76(12):2732–7.

Chapter 4
Role of Oxidative Stress in Hypertension

Sophocles Chrissobolis, Quynh N. Dinh, Grant R. Drummond, and Christopher G. Sobey

Abbreviations

Ang II	Angiotensin II
AT1R	Angiotensin II type 1 receptor
Atox-1	Antioxidant-1
CD40L	CD40 ligand
COX	Cyclooxygenase
CuZnSOD	Copper-zinc superoxide dismutase
DOCA	Deoxycorticosterone acetate
DUOX	Dual oxidase
ECSOD	Extracellular superoxide dismutase
eNOS	Endothelial nitric oxide synthase
G6PD	Glucose-6-phosphate dehydrogenase
GPx	Glutathione peroxidase
ICAM-1	Intracellular adhesion molecule-1
MnSOD	Manganese superoxide dismutase
NO	Nitric oxide
PPARβ	Peroxisome proliferator-activated receptor beta
RNS	Reactive nitrogen species
ROS	Reactive oxygen species
SHR	Spontaneously hypertensive rat
SHRSP	Stroke-prone spontaneously hypertensive rat
siRNA	Small-interfering RNA
SOD	Superoxide dismutase

S. Chrissobolis, Ph.D. (✉) • Q.N. Dinh • G.R. Drummond • C.G. Sobey
Department of Pharmacology, Monash University,
Building 13E, Wellington Rd, Clayton, VIC 3800, Australia
e-mail: Sophocles.Chrissobolis@monash.edu

M. Rodriguez-Porcel et al. (eds.), *Studies on Atherosclerosis*,
Oxidative Stress in Applied Basic Research and Clinical Practice,
DOI 10.1007/978-1-4899-7693-2_4

TNFα Tumour necrosis factor alpha
VCAM-1 Vascular cell adhesion molecule-1
WKY Wistar-Kyoto

Introduction

Oxidative stress refers to an increase in steady-state levels of reactive oxygen species (ROS), including superoxide (the precursor for multiple ROS), and can be due to increased production of superoxide and reactive nitrogen species [RNS], and/or decreased expression or activity of antioxidant enzymes such as superoxide dismutases (SOD's) and glutathione peroxidises (GPx's) that regulate subcellular ROS levels [1, 2]. ROS can have direct effects; however, since superoxide reacts extremely efficiently with nitric oxide (NO) to form the RNS peroxynitrite, ROS can reduce NO bioavailability and adversely affect NO signalling, causing endothelial dysfunction. As such, oxidative stress is involved in the pathogenesis of hypertension [3]. Hypertension is a complex condition and a major risk factor for cardiovascular events [4]. Although many of the cases of hypertension have an unknown cause (defined as essential hypertension) [5], there is much experimental evidence supporting a role for increased ROS in the pathogenesis of hypertension [6].

The purpose of this chapter is to examine the relationship between oxidative stress and hypertension in experimental models, including the involvement of oxidative stress and novel downstream signalling mechanisms. Vascular *N*ADPH *ox*idases (Nox) are by far the most researched source of ROS in hypertension, and are thought to be a predominant underlying cause of oxidative stress in numerous chronic cardiovascular diseases [7], and hence will be a major focus of our discussion. Antioxidant defence mechanisms (i.e. SOD's, GPx) may limit vascular oxidative stress and protect against hypertension and they will be discussed also. Innate immune cells produce ROS to destroy invading pathogens; however, sustained inflammation can lead to oxidative stress. In addition to being a major source of ROS in the vasculature, NADPH oxidases are also a source of ROS in immune cells [7], thus the recent concept linking the immune system, oxidative stress, and hypertension will be examined, and finally we will briefly address clinical data providing an association between oxidative stress and hypertension, in particular the link between genetic abnormalities and oxidative stress in hypertension.

Hypertension: Involvement of the Central Nervous System and Peripheral Mechanisms

Hypertension is a complex clinical condition, with the vast majority of cases classified as essential hypertension, meaning the exact cause is unknown [8, 9]. Based on data from 2007–2010, 33 % of US adults >20 years of age have hypertension [10]. Hypertension is a risk factor for many disease, including stroke and myocardial infarction [11], thus understanding the mechanisms that contribute to it remains an active an important area of research. Alterations in function of the central nervous system, kidneys and the vasculature are all implicated in the pathogenesis of hypertension [11]. In response to an elevation in blood pressure, the kidney responds by increasing urinary sodium excretion, ultimately resulting in a reduction in body fluid volumes and restoring blood pressure to normal [11]. Total peripheral resistance, governed by vascular tone, is also a key determinant of blood pressure. Thus, when the ability of blood vessels to relax is compromised, or their contractile status is enhanced, this may contribute to elevated total peripheral resistance. Sympathetic nervous system activation also contributes to hypertension, as enhanced sympathetic outflow may result in increased total peripheral resistance, cardiac output and sodium retention through adrenergic receptor activation in the vasculature, heart and kidney, respectively. Importantly, oxidative stress in each of these three systems is thought to be involved in hypertension.

Role of Nox Isoforms in Regulating Blood Pressure and Vascular Dysfunction

Vascular NADPH oxidases are each a multisubunit complex composed of a membrane-bound core catalytic subunit (Nox [which consists of isoforms 1–5] and dual oxidase [DUOX] subunits -DUOX1 and DUOX2), and up to five regulatory subunits: p22phox, DUOX activator 1 [DUOXA1] and DUOXA2 which are important in maturation and expression of NOX and DUOX subunits in membranes; p67phox and NOX activator 1 (NOXA1), which are important in enzyme activation; p47phox, NOX organiser 1 (NOXO1) and p40phox, which are each involved in spatial organisation of various components of the enzyme complex. Some Nox isoforms require a small G protein (Rac1 or Rac2) for their activation [7, 12]. NADPH oxidase catalyses the transfer of electrons—with cytosolic NADPH as the electron donor—to molecular oxygen. Electron transfer to molecular oxygen results in superoxide being released from the oxidase [7].

The catalytic and regulatory subunits are known to be expressed throughout the vascular wall, including in adventitial fibroblasts, endothelial cells and vascular smooth muscle cells. Endothelial cells express Nox1, Nox2, Nox4, Nox5, p22phox, p67phox, p47phox, p40phox and Rac1 [7, 13–19]. Vascular smooth muscle cells express Nox1, Nox4 and Nox5 [7, 20–23], whereas adventitial fibroblasts express Nox2 and Nox 4 [24–26].

Baseline Blood Pressure

Overall, Nox's do not seem to markedly influence blood pressure under basal conditions. Baseline blood pressure is reported to be lower in Nox1-deficient mice compared to wild-type controls [27, 28]. These studies suggest some contribution of

Nox1 to baseline blood pressure control, although another study suggested no effect of Nox1-deficiency on baseline pressure [29]. Overexpression of Nox1 in vascular smooth muscle cells had no effect on blood pressure [30]. Baseline blood pressure was not influenced by Nox2 deficiency [27, 31], although one study reported a slight reduction in Nox2-deficient mice [32]. Endothelial cell Nox2 overexpression had no effect on baseline blood pressure [33, 34]. In contrast to Nox1, Nox4 may have a beneficial effect on blood pressure, since mice overexpressing Nox4 in endothelial cells had lower blood pressure than wild-type controls [35].

Commonly Used Models of Experimental Hypertension

Ang II is commonly administered to rodents to produce and study hypertension and its associated organ dysfunction. Consequently, a vast array of literature exists to suggest that Ang II-induced hypertension is associated with oxidative stress and endothelial dysfunction and has been reviewed elsewhere [1, 6, 36–40]. The role of Nox in mediating hypertension and associated vascular effects will be discussed below, as will the involvement of Nox in mediating hypertension and vascular dysfunction in mineralocorticoid-dependent hypertension (in the section "Mineralocorticoid-Dependent Hypertension"). Information on the contribution of oxidative stress to hypertension in SHR is also important given this animal model is regarded as a model of essential hypertension. The role of Nox in hypertension in SHR is discussed below (in the section "SHR").

Ang II-Dependent Hypertension

Ang II-induced elevation in blood pressure and vascular superoxide was markedly attenuated in Nox1-deficient vs wild-type mice, as was Ang II-induced endothelial dysfunction [29]. Interestingly, Nox1 does not appear to be involved in the early phase of blood pressure elevation, but rather appears to play a role in sustained elevations of blood pressure [28, 29]. Consistent with this, the blood pressure response to Ang II was significantly enhanced in mice overexpressing Nox1 in vascular smooth muscle cells (Nox1 Tg) when compared with control mice [30, 41]. Treatment of mice with the superoxide spin trap agent tempol reduced the Ang II response in both groups of mice, with a much larger reduction in the transgenic mice, suggesting the enhanced pressor response was superoxide-mediated [30]. Ang II-induced vascular superoxide production, endothelial dysfunction, and reduced NO bioavailability was also much greater in Nox1-Tg compared to control mice [30, 41]. Nox1 may also mediate vascular hypertrophy, as enhanced aortic hypertrophy in transgenic mice compared to littermate controls was partially reversed by tempol [30], and Ang II-induced medial hypertrophy was reportedly Nox1-dependent [28]. However, another study reported no role for Nox1 in Ang II-induced vascular hypertrophy [29].

Ang II-induced vascular hypertrophy, oxidative stress and endothelial dysfunction were all absent in Nox 2-deficient mice treated with Ang II [27, 31, 42].

However, Nox2-deficiency does not appear to typically attenuate the pressor effect of Ang II [27, 31, 42], suggesting Nox2 mediates the vascular, but not the pressor actions of Ang II. However, a very recent study reported partial attenuation (~50 %) of the pressor effect of Ang II in Nox2-deficient mice [43]. These differences may be related to different doses used in those studies. Nevertheless, Ang II-induced increases in vascular oxidative stress and blood pressure were attenuated by gp91ds-tat—an inhibitor of the interaction of p47phox and Nox2 [44]. Apocynin, an NADPH oxidase inhibitor which inhibits association of p47phox with the membrane bound heterodimer [7], was also able to attenuate Ang II-induced hypertension in association with inhibition of vascular oxidative stress and endothelial dysfunction [45]. In mice overexpressing Nox2 in endothelial cells (Nox2-Tg), endothelium-dependent relaxation was similar when compared with wild-type mice [33, 34]. Moreover, a low dose of Ang II which increased blood pressure in these mice was without effect in wild-type mice, an effect associated with potentiation of Ang II-induced increases in vascular endothelial superoxide in Nox2-Tg mice [33]. Thus, endothelial cell Nox2 may be an important mediator of the Ang II-induced increases in oxidative stress and increased blood pressure.

Nox4 (an inducible Nox isoform) reportedly protects the vasculature from Ang II, whereby vascular endothelial dysfunction, hypertrophy and inflammation in response to Ang II was augmented in Nox4-deficient vs control mice [46]. Consistent with this, endothelium-dependent relaxation was enhanced in mice overexpressing Nox4 in endothelial cells. This effect is H_2O_2-dependent since it was prevented by catalase [35]. The initial increase in blood pressure in response to Ang II was also blunted in endothelial Nox4 overexpressing mice compared to their controls [35].

Mediators of Nox Signalling in Ang II-Dependent Hypertension

Research has also been carried out to characterise Nox-mediated signalling leading to the pressor and vascular effects of Ang II. Vascular oxidative stress and hypertension in response to Ang II were attenuated in p47phox-deficient vs wild-type mice [47]. In that study, NADPH oxidase was identified as the source of oxidative stress in endothelial cells since p47phox deficiency in cultured endothelial cells resulted in blunted Ang II-induced superoxide production [47]. Ang II-induced translocation of p47phox to the plasma membrane (which is necessary for Nox2 activation) was blunted in mice deficient in mitogen-activated protein kinase-activated protein kinase 2. This was also associated with blunted Ang II-induced hypertension, vascular superoxide production and NADPH oxidase activity, and inflammation [48]. NADPH is required for the activity of NADPH oxidase, and glucose-6-phosphate-dehydrogenase (G6PD) is the rate-limiting enzyme in the pentose phosphate pathway that generates two molecules of NADPH from $NADP^+$ through oxidation of G6P. Vascular NADPH content was reduced in G6PD-deficient vs wild-type mice, and Ang II-induced vascular oxidative stress, medial hypertrophy and increased blood pressure were blunted in G6PD-deficient mice, suggesting that G6PD, by limiting production of the substrate for NADPH oxidase, mediates the vascular and pressor effects of Ang II [49].

Evidence for an involvement of the Ang II type 1 receptor (AT1R) upstream of NADPH oxidase activation has been reported in that increased blood pressure in response to Ang II was associated with increased vascular expression of p47phox and Nox2, effects that were inhibited with the AT1R antagonist losartan [50]. Ang II-induced increases in blood pressure were attenuated in rats administered small-interfering RNA (siRNA) sequences to p22phox, which was associated with reduced renal NADPH oxidase activity, Nox expression and oxidative stress [51].

Mineralocorticoid-Dependent Hypertension

The involvement of Nox in hypertension, vascular oxidative stress and dysfunction has also been investigated in a model of mineralocorticoid-dependent hypertension. Deoxycorticosterone acetate (DOCA)/salt-induced hypertension and vascular oxidative stress were virtually abolished in gp91phox-deficient mice [52]. Similarly, the initial rise (i.e. within 2–3 days) in blood pressure in response to DOCA/salt treatment was attenuated in gp91phox-and p47phox-deficient mice, although there was no difference at later time points [32]. Interestingly, despite the absence of hypertension in aldosterone-treated mice, cerebral vascular oxidative stress and endothelial dysfunction was abolished in Nox2-deficient mice [53].

Studies using apocynin have reported similar effects to those using Nox2-deficient mice. Apocynin attenuated DOCA/salt-induced hypertension [54–57] in association with reduced vascular oxidative stress [56, 57], NADPH oxidase activity [56], p22phox expression [55], p47phox expression [57] and endothelial dysfunction [56]. Vascular stiffness, also a feature of hypertension, which was elevated in aorta from DOCA/salt-treated vs control mice, was also attenuated by apocynin. Increased aortic medial thickness and intima-media thickness of carotid arteries in DOCA/salt-treated vs control mice was also prevented by apocynin [57].

Epicatechin, a flavonol—which reduces vascular NADPH oxidase activity, p22phox and p47phox expression during DOCA/salt-induced hypertension - also reduced hypertension, vascular oxidative stress and endothelial dysfunction [58]. Red-wine polyphenols had similar effects, suggesting flavonols provide anti-hypertensive and vascular protective effects in mineralocorticoid-dependent hypertension due to their ability to inhibit NADPH oxidase activity [56]. In addition to its effects on NADPH oxidase activity, apocynin (in combination with tetrahydrobiopterin) can reportedly recouple endothelial nitric oxide synthase (eNOS) to restore NO generation in DOCA/salt-induced hypertensive mice [59].

SHR

In stroke prone SHR (SHRSP) fed a high-salt diet, tempol pre-treatment prevented a further increase in blood pressure, an effect associated with decreased vascular superoxide, thus providing evidence that oxidative stress is associated with hypertension in this model [60]. Nox1, Nox2 and Nox4 mRNA and protein expression, and NADPH oxidase activity were elevated in mesenteric resistance artery vascular smooth muscle cells from SHR when compared to Wistar Kyoto (WKY) control rats. NADPH oxidase activity in vascular smooth muscle cells from WKY rats or SHR was decreased by the Nox 1/4 inhibitor GKT136901 [61]. In cerebral arteries, Nox4 mRNA was increased in SHR vs WKY rats in association with elevated NADPH oxidase activity, with no reported increase in Nox1, Nox4, p22phox and p47phox expression [62]. Vascular NADPH oxidase activity, which was increased in aortic homogenates from SHR vs WKY rats, was attenuated by mito-tempol (a scavenger of mitochondrial ROS) and celecoxib (a selective cyclooxygenase [COX]-2 inhibitor), and celecoxib also attenuated increased superoxide levels and Nox1 and Nox4 mRNA expression [45]. These results suggest that under some conditions mitochondrial-derived ROS and COX may activate NADPH oxidase.

Apocynin improved endothelium-dependent relaxation and reversed oxidative stress in coronary arteries of SHR [63]. A recent study reported that pressure-induced constriction (i.e. myogenic response) of afferent arterioles was enhanced in SHR vs WKY rats, and both tempol and gp91ds-tat attenuated the pressure-induced constriction in SHR but not WKY rats, suggesting that Nox2-derived superoxide contributes to the enhanced myogenic response [64]. In SHR (but not WKY rats) the peroxisome proliferator-activated receptor beta (PPARβ) activator GW0742 elicited an anti-hypertensive effect. Furthermore, in aorta isolated from SHR, superoxide levels, NADPH oxidase activity, p22phox and p47phox mRNA expression was increased relative to WKY rats, and this increase was abolished in SHR treated with the PPARβ activator GW0742. Such results indicate the anti-hypertensive effect of PPARβ activation may partly involve inhibition of NADPH oxidase-dependent ROS generation [65].

Vascular expression of Nox1 and Nox2, but not Nox4 was increased in aged SHR vs WKY rats [66], whereas Nox4, but not Nox1 or Nox2 expression was increased in aged vs young rats [67]. Oxidative stress and endothelial dysfunction in aged rats was reversed by apocynin [66, 67]. Thus, involvement of Nox isoforms in oxidative stress may differ in aging depending on the presence of hypertension. Furthermore, in the SHR, the AT1R antagonist candesartan inhibited the increased medial thickness, fibrosis and oxidative stress in the vasculature, effects associated with a reduction in blood pressure and reduced vascular expression of Nox1 and Nox4 [68].

Role of Nox in the Brain During Hypertension

An important demonstration that ROS generation in the central nervous system is crucial in hypertension (Fig. 4.1) was the finding that Ang II-induced hypertension is associated with elevations in superoxide production in the subfornical organ—a brain region lying outside the blood-brain barrier and a sensor of blood-borne Ang II [69]. Intracerebroventricular administration of Ang II increased blood pressure, and adenoviral-mediated delivery of Nox2 and Nox4 siRNA to the subfornical organ of the brain individually partly prevented the blood pressure increase, whereas the combination of both abolished the increase [70]. Further evidence for a role of NADPH

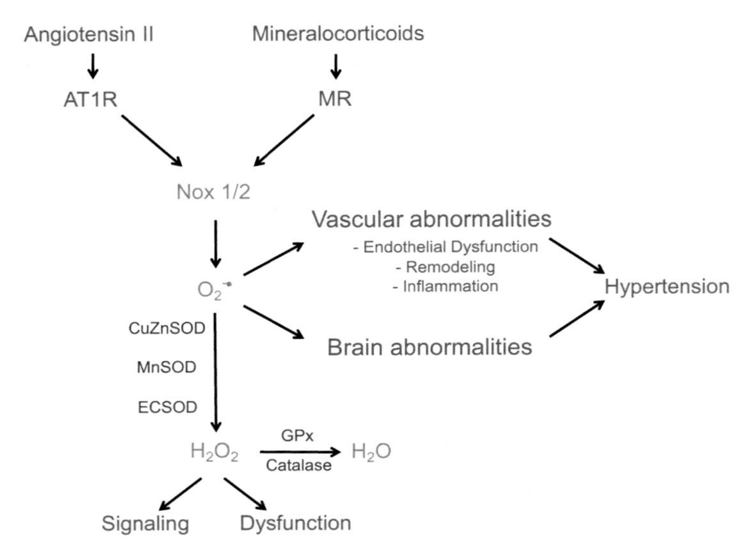

Fig. 4.1 Hypertensive stimuli (e.g. angiotensin II, mineralocorticoids) increase activity of Nox 1 and Nox2 oxidase. Superoxide produced via Nox1 and Nox2 oxidase promotes hypertension through vascular abnormalities including endothelial dysfunction, remodelling and inflammation. Oxidative stress in other organs (e.g. brain) may also promote hypertension. SOD's (shown in *blue*) are antioxidant enzymes that scavenge superoxide, resulting in protection against vascular and brain abnormalities and thus hypertension. Superoxide is converted into H_2O_2 by SOD's. H_2O_2 is an important signalling molecule, but can also form the hydroxyl radical, a highly reactive ROS that can cause cellular dysfunction. GPx and catalase, which can degrade H_2O_2, protect against angiotensin II-induced vascular abnormalities and hypertension (not shown in diagram, see text for details)

oxidase in the central actions of Ang II was reported recently by Lob and colleagues, where deletion of p22phox in the subfornical organ resulted in marked reduction of both the hypertensive response and vascular inflammatory response to Ang II [71]. Furthermore, inhibition of Rac1 by transfecting adenovirus vectors encoding dominant-negative Rac1 into the nucleus tractus solitarius decreased blood pressure in SHRSP, but not WKY rats, which was associated with reduced NADPH oxidase activity and ROS generation [72].

Role of Antioxidant Enzymes in Regulating Blood Pressure and Vascular Function

As mentioned earlier, oxidative stress can result from diminished activity of antioxidant enzymes such as SOD's. There are three isoforms of SOD expressed in the vasculature: cytosolic or copper-zinc SOD (CuZnSOD or SOD-1), manganese SOD (MnSOD or SOD-2) localised in mitochondria, and an extracellular form of

CuZnSOD (ECSOD or SOD-3) [73, 74]. Although they are a product of different genes and their subcellular localization is different, they all catalyse the same reaction—the dismutation of superoxide to hydrogen peroxide and oxygen [74]. Their roles in protecting against hypertension, vascular oxidative stress and dysfunction are discussed below.

CuZnSOD

CuZnSOD is the major intracellular SOD and it exists as a 32 kDa homodimer. It is mainly localised in the cytosol, with some expression reported in the intermembrane space of mitochondria [74].

Baseline blood pressure in homozygous CuZnSOD-deficient (CuZnSOD$^{-/-}$) mice (i.e. mice in which both copies of the gene have been deleted) appeared to be lower compared to wild-type controls, despite lower total SOD activity, impaired vascular endothelial function and elevated superoxide levels [75, 76]. Thus, these data suggest that oxidative stress may not necessarily promote hypertension during CuZnSOD-deficiency. In heterozygous CuZnSOD-deficient (CuZnSOD$^{+/-}$) mice (i.e. mice in which one copy of the gene has been deleted), baseline blood pressure was similar to wild-type mice, and although SOD activity was decreased, superoxide levels and endothelial function were unaltered. Thus, blood pressure and vascular changes appear to be observed only in conditions of homozygous CuZnSOD deficiency. During aging, although blood pressure was unaltered, SOD activity was decreased, superoxide levels were increased and endothelial function (which could be restored following treatment with tempol) was impaired [77].

Overexpression of human SOD (which leads to increased vascular, brain and cardiac CuZnSOD activity) partially inhibited Ang II-induced vascular superoxide production and hypertension, although Ang II-induced vascular hypertrophy was unaffected [78]. Overexpression of CuZnSOD in the nucleus tractus solitarius decreased blood pressure in SHRSP [72]. Furthermore, adenoviral-mediated delivery of SOD cytoplasmically targeted to the subfornical organ prevented Ang II-induced ROS production and hypertension [69]. Thus, while CuZnSOD does not appear to result in lower blood pressure under normal conditions, it appears to protect against hypertension in the SHR and in response to Ang II in association with reduced oxidative stress.

MnSOD

MnSOD is a manganese-containing enzyme composed of a 96 kDa homotetramer and is localised in the mitochondrial matrix. It is involved in the dismutation of superoxide generated by enzymes of the mitochondrial respiratory chain [74]. The importance of MnSOD is underscored by the observation that mice completely deficient in

MnSOD die within a few weeks after birth [79], suggesting mitochondrial-mediated oxidative stress is a crucial antioxidant defence mechanism for long-term survival. MnSOD expression in the vasculature is increased in hypertension [80–82].

Baseline blood pressure in aged (i.e. 2-year-old) mice was increased in MnSOD-deficient vs wild-type control mice. In adult (i.e. 6 month old) mice, although baseline blood pressure was not different, a high-salt diet increased blood pressure in MnSOD-deficient, but not wild-type mice [83]. These findings implicate a role for MnSOD-deficiency as a protective mechanism in response to certain hypertensive stimuli, and during aging. MnSOD-deficiency had no effect on blood pressure following treatment with a sub-pressor dose of Ang II, although Ang II-induced endothelial dysfunction was exacerbated by MnSOD deficiency, suggesting that MnSOD protects the vasculature in response to elevated Ang II [84].

Ang II is known to increase superoxide formation in vascular cells, and this may occur via several mechanisms, including increased formation in mitochondria [85]. Following hypertension established by Ang II [86, 87] or DOCA/salt [87] administration, mito-tempol partially reversed the hypertension. Another study reported pharmacological inhibition of mitochondrial oxidative stress markedly suppressed DOCA/salt-induced hypertension, suggesting mitochondria are a major source of ROS production in DOCA/salt-induced hypertension [32]. Furthermore, mice overexpressing MnSOD demonstrated attenuated Ang II-induced hypertension and vascular oxidative stress [87]. Mito-tempol also reversed Ang II-induced vascular oxidative stress and endothelial dysfunction, and the reduction in NO bioavailability by Ang II and DOCA/salt [87]. Taken together, these findings demonstrate an important role for mitochondrial ROS in contributing to vascular dysfunction and hypertension. Mitochondrial ROS formation reportedly contributes to age-dependent endothelial dysfunction [88].

Most recently, evidence for an interaction between mitochondrial ROS and Nox2 was reported, where Ang II-induced increases in endothelial cell superoxide were inhibited by mito-tempol. Furthermore, diazoxide (which stimulates mitochondrial superoxide)-induced superoxide production in the cytoplasm was inhibited by gp91ds-tat and Nox2 siRNA, but not Nox1, Nox4 or Nox5 siRNA [86]. Ang II-induced increases in vascular mitochondrial superoxide were also inhibited in Nox2-deficient mice and by gp91ds-tat [43]. A further report found that Nox2 inhibition with apocynin impaired endothelial function in MnSOD-deficient mice [89]. These authors suggested that the functional consequences of increased ROS may depend on mitochondrial antioxidant status [89].

ECSOD

ECSOD is a CuZn-containing SOD and is the major SOD in the extracellular space of blood vessels. It exists as a 135 kDa homotetramer composed of two disulfide-linked dimers in most species, and is highly expressed in blood vessels, particularly smooth muscle cells [74]. Baseline blood pressure was unaltered by ECSOD deficiency [90, 91], whereas Ang II-induced and renovascular (2 kidney-1 clip) hypertension was exacerbated in ECSOD-deficient mice compared to wild-type controls [90, 91]. ECSOD deficiency had no effect on L-NAME-induced hypertension [91]. Thus, ECSOD may protect against some, but not all forms of hypertension. Protection by ECSOD in Ang II-induced hypertension was associated with enhanced superoxide production and reduced endothelium-dependent vasodilatation in Ang II-treated ECSOD-deficient vs wild-type mice in mesenteric resistance vessels [90]. In that study in ECSOD-deficient mice, Ang II reduced aortic oxidative stress, improved aortic endothelial function, and reduced NADPH oxidase activity. Thus in resistance vessels ECSOD protected against Ang II-induced detrimental vascular effects and thus hypertension, whereas in large vessels other mechanisms appear to compensate during ECSOD deficiency to preserve their function and hypertension [90]. Recombinant ECSOD administered to hypertensive ECSOD-deficient mice was able to reduce blood pressure and restore impaired NO bioavailability and endothelial function [91].

Subsequent work has sought to examine the cell types involved and the mechanism of protection by ECSOD. ECSOD is a secretory copper-containing enzyme, and requires the full activity of the copper transporter Menkes ATPase. Ang II-induced hypertension, vascular superoxide production and endothelial dysfunction were enhanced in Menkes ATPase-deficient mice vs wild-type controls. Tempol partially inhibited Ang II-induced hypertension in Menkes ATPase-deficient mice, suggesting the exacerbated pressor response is due to elevated levels of superoxide [92]. ECSOD activity also requires the copper transport protein Antioxidant-1 (Atox-1), which likely interacts with Menkes ATPase to deliver copper to the secretory copper enzymes. Ang II-induced hypertension, vascular superoxide production and endothelial dysfunction were all enhanced in Atox-1-deficent mice, and these effects were reversed by tempol [93]. Thus, Menkes ATPase and Atox-1 are involved in protecting against the vascular and pressor actions of Ang II, most likely through regulation of ECSOD activity, and subsequently oxidative stress. Specific deletion of ECSOD in vascular smooth muscle cells did not augment the pressor response to Ang II or Ang II-induced vascular inflammation, whereas deletion of ECSOD in circumventricular organs resulted in elevated blood pressure in response to an otherwise non-pressor dose of Ang II [94], highlighting the importance of modulation of oxidative stress by ECSOD in the brain as a potential protective mechanism during hypertension.

Glutathione Peroxidases

Glutathione peroxidases (GPxs) are a family of enzymes homologous to the selenocysteine-containing mammalian GPx-1 that uses reduced glutathione as a substrate in the reduction of hydrogen peroxide to water, with oxidised glutathione formed in the process [95]. GPx-1 is one of the most abundant members of the GPx enzyme family, expressed in the cytosol and mitochondria of all cells, and in peroxisomal compartments in some cells [95], and as such is important in preventing detrimental accumulation of intracellular hydrogen peroxide. GPx-2 is an epithelial-specific enzyme highly expressed in intestine, GPx-3 is a secreted subtype; however, GPx-3 mRNA has been detected in blood vessels [96]. GPx-4 is widely expressed and differs in its substrate specificity compared to other family members [95].

GPx-1 deficiency (GPx-1$^{-/-}$) had no effect on baseline blood pressure or Ang II-induced hypertension, and had no effect on wall:lumen ratio or cross-sectional area under basal conditions or in response to Ang II; however, it accelerated Ang II-induced cardiac hypertrophy and dysfunction [97]. Endothelium-dependent relaxation of mesenteric arterioles was significantly impaired in GPx-1$^{+/-}$ mice, which was associated with vascular oxidative stress, perivascular matrix deposition, intimal thickening, and an increased number of adventitial fibroblasts [98]. GPx-1 deficiency also resulted in carotid artery endothelial dysfunction in GPx-1$^{-/-}$ mice. In GPx-1$^{+/-}$ mice, treatment of vessels with Ang II (at a concentration that had no effect in wild-type mice) resulted in endothelial dysfunction. Complementary experiments showed that a higher concentration of Ang II caused endothelial dysfunction in wild-type mice, but had no effect in mice overexpressing GPx-1 [99]. These data suggest that GPx-1 protects against Ang II-induced endothelial dysfunction. In human microvascular endothelial cells, GPx-1 deficiency-enhanced intracellular adhesion molecule (ICAM)-1 and vascular cell adhesion molecule (VCAM)-1 expression, and GPx-1 suppression enhanced tumour necrosis factor alpha (TNFα)-induced ROS production and ICAM-1 expression, suggesting GPx-1 protects against vascular proinflammatory redox signalling [95]. Thus, GPx-1 protects against vascular oxidative stress, inflammation and endothelial dysfunction, although it may not protect against hypertension.

GPx-1 deficiency (GPx-1$^{-/-}$) resulted in vascular oxidative and nitrosative stress, eNOS dysfunction, as well as endothelial dysfunction in aged mice compared to age-matched controls. The enhanced oxidative stress may be localised to adventitia, as levels of adventitial ROS production were enhanced during aging. Moreover, increased vascular leukocyte infiltration was reported in aged GPx-1$^{-/-}$ vs control mice [100]. Although the effects of GPx-1 deficiency on hypertension were not examined in this study, GPx-1 deficiency during aging results in exacerbated vascular oxidative stress, endothelial dysfunction and inflammation during aging. Dietary supplementation with selenium, which results in increased GPx-1 activity, resulted in reduced wall thickness of the left coronary artery in SHR, whereas selenium deficiency resulted in increased wall thickness of the abdominal aorta [101], suggesting that targeting GPx-1 to increase its activity may confer vascular protection during hypertension.

Recent Concepts: Immunity, Oxidative Stress and Hypertension

Data from both experimental models and clinical studies have implicated the involvement of immune cells in hypertension [9]. A potential major advance in understanding mechanisms contributing to hypertension was reported recently, whereby mice lacking T and B lymphocytes ($RAG^{-/-}$ mice) failed to develop hypertension in response to Ang II and DOCA-salt. This was associated with a reduction in vascular oxidative stress and endothelial dysfunction in response to Ang II. In $RAG^{-/-}$ mice, the pressor effect, oxidative stress and vascular dysfunction in response to Ang II could be restored by replacement of T cells, implicating a role for T cells in Ang II-induced hypertension and vascular dysfunction [102]. As mentioned earlier, in addition to being a major source of ROS in the vasculature, NADPH oxidase is a major source of ROS in immune cells [7]. Adoptive transfer of T cells that lack p47phox was unable to restore the Ang II-induced increase in superoxide, and only resulted in partial restoration of the pressor response to Ang II, implicating p47phox-containing NADPH oxidase expressed on T cells in Ang II-induced oxidative stress and hypertension. The importance of NADPH oxidase expressed on T cells is consistent with increased T cell expression of Nox2, p22phox, p47phox and p67phox in response to Ang II [102]. This finding was confirmed by a recent study, which reported that adoptive transfer of T cells from male mice was able to restore the pressor effect of Ang II in $RAG^{-/-}$ male mice [103]. Interestingly, however, this phenomenon appears to be sex-specific, as T cells isolated from female mice were not able to restore the hypertensive response to Ang II when they were adoptively transferred into male $RAG^{-/-}$ mice [103].

CD40 ligand (CD40L), which is important in vascular infiltration of immune cells, is an activator of the innate and adaptive immune system [104]. Ang II-induced vascular dysfunction, increases in vascular ROS, Nox2 and p47phox expression were blunted in CD40L-deficient mice, suggesting an important role for CD40L in the detrimental vascular actions of Ang II. Selective depletion of macrophages also resulted in reduced Ang II-induced vascular oxidative stress in association with reduced hypertension and restoration of endothelial function [105]. Adoptive transfer of monocytes into mice with selective depletion of macrophages restored the Ang II-induced increase in blood pressure and endothelial dysfunction. However, adoptive transfer of monocytes from gp91phox-deficient mice did not impact systolic blood pressure, endothelial function, vascular oxidative stress and inflammation, suggesting monocytes need to contain Nox2 to mediate Ang II-induced hypertension, vascular dysfunction, oxidative stress and inflammation [105].

Clinical Links Between Oxidative Stress and Hypertension

Clinical evidence exists to support an association between oxidative stress and hypertension, although a cause and effect relationship between oxidative stress and blood pressure is more difficult to establish. Such data relies heavily on indirect evidence, such as increased plasma markers of oxidative stress in patients with hypertension [106].

In the vasculature, Nox5 expression was increased in hypertensive patients with coronary artery disease [107]. Ang II-induced oxidative stress was enhanced in peripheral resistance arteries of hypertensive vs normotensive patients [108]. The importance of Nox in cardiovascular disease is further underscored by genetic studies in humans. The p22phox gene -930 (A/G) polymorphism determines expression of p22phox and expression of NADPH oxidase in phagocytic cells from patients with essential hypertension [109]. In hypertensive patients, the T allele of the p22phox C242T polymorphisms is associated with increased NADPH oxidase activity and left ventricular mass/height, suggesting a role for Nox in cardiac remodelling during hypertension [110]. The C242T CYBA (i.e. the human gene that encodes p22phox) polymorphism is associated with essential hypertension, and hypertensives carrying the CC genotype of this polymorphism exhibit features of NADPH oxidase-mediated oxidative stress and endothelial damage [111]. The -675 (A/T) CYBA polymorphism is also associated with essential hypertension, and the prevalence of the TT genotype was higher in hypertensives than normotensives. TT subjects display features of NADPH oxidase-mediated oxidative stress in phagocytic cells [112].

Conclusion

A wealth of information exists to support the concept that oxidative stress contributes to hypertension (Fig. 4.1), primarily from genetic studies reporting attenuated pressor responses to commonly used experimental stimuli such as Ang II. Of the potential ROS sources, vascular NADPH oxidases appear to be commonly associated with hypertension, with recent studies highlighting the importance of cell-specific Nox subunits and their contribution to baseline blood pressure, and in response to pressor agents, making them attractive targets for development of therapeutics. While there is a clear need for further research, other molecular targets—such as subcellular-specific SOD's, and GPx-1 - may also provide targets for novel drug development to suppress oxidative stress once it has developed. Exciting new research into the contribution of the immune system to hypertension may provide avenues for the development of therapeutic agents that target ROS-producing immune cells to treat hypertension [9].

References

1. Chrissobolis S, Miller AA, Drummond GR, Kemp-Harper BK, Sobey CG. Oxidative stress and endothelial dysfunction in cerebrovascular disease. Front Biosci. 2011;16:1733–45.
2. Chrissobolis S, Faraci FM. The role of oxidative stress and NADPH oxidase in cerebrovascular disease. Trends Mol Med. 2008;14(11):495–502.
3. Touyz RM. Reactive oxygen species, vascular oxidative stress, and redox signaling in hypertension: what is the clinical significance? Hypertension. 2004;44(3):248–52.
4. Ventura HO, Taler SJ, Strobeck JE. Hypertension as a hemodynamic disease: the role of impedance cardiography in diagnostic, prognostic, and therapeutic decision making. Am J Hypertens. 2005;18(2 Pt 2):26S–43.

5. Messerli FH, Williams B, Ritz E. Essential hypertension. Lancet. 2007;370(9587):591–603.
6. Montezano AC, Touyz RM. Reactive oxygen species, vascular Noxs, and hypertension: focus on translational and clinical research. Antioxid Redox Signal. 2014;20(1):164–82.
7. Drummond GR, Selemidis S, Griendling KK, Sobey CG. Combating oxidative stress in vascular disease: NADPH oxidases as therapeutic targets. Nat Rev Drug Discov. 2011;10(6): 453–71.
8. Coffman TM. Under pressure: the search for the essential mechanisms of hypertension. Nat Med. 2011;17(11):1402–9.
9. Dinh QN, Drummond GR, Sobey CG, Chrissobolis S. Roles of inflammation, oxidative stress, and vascular dysfunction in hypertension. Biomed Res Int. 2014;2014:406960.
10. Go AS, Mozaffarian D, Roger VL, Benjamin EJ, Berry JD, Blaha MJ, et al. Heart disease and stroke statistics--2014 update: a report from the American Heart Association. Circulation. 2014;129(3):e28–e292.
11. McMaster WG, Kirabo A, Madhur MS, Harrison DG. Inflammation, immunity, and hypertensive end-organ damage. Circ Res. 2015;116(6):1022–33.
12. Bedard K, Krause KH. The NOX family of ROS-generating NADPH oxidases: physiology and pathophysiology. Physiol Rev. 2007;87(1):245–313.
13. Drummond GR, Sobey CG. Endothelial NADPH oxidases: which NOX to target in vascular disease? Trends Endocrinol Metab. 2014;25(9):452–63.
14. Brandes RP, Weissmann N, Schroder K. NADPH oxidases in cardiovascular disease. Free Radic Biol Med. 2010;49(5):687–706.
15. Selemidis S, Sobey CG, Wingler K, Schmidt HH, Drummond GR. NADPH oxidases in the vasculature: molecular features, roles in disease and pharmacological inhibition. Pharmacol Ther. 2008;120(3):254–91.
16. Ago T, Kitazono T, Kuroda J, Kumai Y, Kamouchi M, Ooboshi H, et al. NAD(P)H oxidases in rat basilar arterial endothelial cells. Stroke. 2005;36(5):1040–6.
17. Dworakowski R, Alom-Ruiz SP, Shah AM. NADPH oxidase-derived reactive oxygen species in the regulation of endothelial phenotype. Pharmacol Rep. 2008;60(1):21–8.
18. Sorescu D, Weiss D, Lassegue B, Clempus RE, Szocs K, Sorescu GP, et al. Superoxide production and expression of nox family proteins in human atherosclerosis. Circulation. 2002;105(12):1429–35.
19. BelAiba RS, Djordjevic T, Petry A, Diemer K, Bonello S, Banfi B, et al. NOX5 variants are functionally active in endothelial cells. Free Radic Biol Med. 2007;42(4):446–59.
20. Ellmark SH, Dusting GJ, Fui MN, Guzzo-Pernell N, Drummond GR. The contribution of Nox4 to NADPH oxidase activity in mouse vascular smooth muscle. Cardiovasc Res. 2005;65(2):495–504.
21. Lassegue B, Griendling KK. NADPH oxidases: functions and pathologies in the vasculature. Arterioscler Thromb Vasc Biol. 2010;30(4):653–61.
22. Lassegue B, Sorescu D, Szocs K, Yin Q, Akers M, Zhang Y, et al. Novel gp91(phox) homologues in vascular smooth muscle cells : nox1 mediates angiotensin II-induced superoxide formation and redox-sensitive signaling pathways. Circ Res. 2001;88(9):888–94.
23. Moe KT, Aulia S, Jiang F, Chua YL, Koh TH, Wong MC, et al. Differential upregulation of Nox homologues of NADPH oxidase by tumor necrosis factor-alpha in human aortic smooth muscle and embryonic kidney cells. J Cell Mol Med. 2006;10(1):231–9.
24. Chamseddine AH, Miller Jr FJ. Gp91phox contributes to NADPH oxidase activity in aortic fibroblasts but not smooth muscle cells. Am J Physiol Heart Circ Physiol. 2003;285(6): H2284–9.
25. Haurani MJ, Pagano PJ. Adventitial fibroblast reactive oxygen species as autacrine and paracrine mediators of remodeling: bellwether for vascular disease? Cardiovasc Res. 2007;75(4):679–89.
26. Pagano PJ, Clark JK, Cifuentes-Pagano ME, Clark SM, Callis GM, Quinn MT. Localization of a constitutively active, phagocyte-like NADPH oxidase in rabbit aortic adventitia: enhancement by angiotensin II. Proc Natl Acad Sci U S A. 1997;94(26):14483–8.

27. Chrissobolis S, Banfi B, Sobey CG, Faraci FM. Role of Nox isoforms in angiotensin II-induced oxidative stress and endothelial dysfunction in brain. J Appl Physiol. 2012; 113(2):184–91.
28. Gavazzi G, Banfi B, Deffert C, Fiette L, Schappi M, Herrmann F, et al. Decreased blood pressure in NOX1-deficient mice. FEBS Lett. 2006;580(2):497–504.
29. Matsuno K, Yamada H, Iwata K, Jin D, Katsuyama M, Matsuki M, et al. Nox1 is involved in angiotensin II-mediated hypertension: a study in Nox1-deficient mice. Circulation. 2005; 112(17):2677–85.
30. Dikalova A, Clempus R, Lassegue B, Cheng G, McCoy J, Dikalov S, et al. Nox1 overexpression potentiates angiotensin II-induced hypertension and vascular smooth muscle hypertrophy in transgenic mice. Circulation. 2005;112(17):2668–76.
31. Wang HD, Xu S, Johns DG, Du Y, Quinn MT, Cayatte AJ, et al. Role of NADPH oxidase in the vascular hypertrophic and oxidative stress response to angiotensin II in mice. Circ Res. 2001;88(9):947–53.
32. Zhang A, Jia Z, Wang N, Tidwell TJ, Yang T. Relative contributions of mitochondria and NADPH oxidase to deoxycorticosterone acetate-salt hypertension in mice. Kidney Int. 2011;80(1):51–60.
33. Bendall JK, Rinze R, Adlam D, Tatham AL, de Bono J, Wilson N, et al. Endothelial Nox2 overexpression potentiates vascular oxidative stress and hemodynamic response to angiotensin II: studies in endothelial-targeted Nox2 transgenic mice. Circ Res. 2007;100(7): 1016–25.
34. Murdoch CE, Alom-Ruiz SP, Wang M, Zhang M, Walker S, Yu B, et al. Role of endothelial Nox2 NADPH oxidase in angiotensin II-induced hypertension and vasomotor dysfunction. Basic Res Cardiol. 2011;106(4):527–38.
35. Ray R, Murdoch CE, Wang M, Santos CX, Zhang M, Alom-Ruiz S, et al. Endothelial Nox4 NADPH oxidase enhances vasodilatation and reduces blood pressure in vivo. Arterioscler Thromb Vasc Biol. 2011;31(6):1368–76.
36. Garrido AM, Griendling KK. NADPH oxidases and angiotensin II receptor signaling. Mol Cell Endocrinol. 2009;302(2):148–58.
37. Gori T, Munzel T. Oxidative stress and endothelial dysfunction: therapeutic implications. Ann Med. 2011;43(4):259–72.
38. Guzik TJ, Harrison DG. Vascular NADPH oxidases as drug targets for novel antioxidant strategies. Drug Discov Today. 2006;11(11-12):524–33.
39. Schulz E, Gori T, Munzel T. Oxidative stress and endothelial dysfunction in hypertension. Hypertens Res. 2011;34(6):665–73.
40. Virdis A, Duranti E, Taddei S. Oxidative stress and vascular damage in hypertension: role of angiotensin II. Int J Hypertens. 2011;2011:916310.
41. Dikalova AE, Gongora MC, Harrison DG, Lambeth JD, Dikalov S, Griendling KK. Upregulation of Nox1 in vascular smooth muscle leads to impaired endothelium-dependent relaxation via eNOS uncoupling. Am J Physiol Heart Circ Physiol. 2010;299(3):H673–9.
42. Girouard H, Park L, Anrather J, Zhou P, Iadecola C. Angiotensin II attenuates endothelium-dependent responses in the cerebral microcirculation through nox-2-derived radicals. Arterioscler Thromb Vasc Biol. 2006;26(4):826–32.
43. Dikalov SI, Nazarewicz RR, Bikineyeva A, Hilenski L, Lassegue B, Griendling KK, et al. Nox2-induced production of mitochondrial superoxide in angiotensin II-mediated endothelial oxidative stress and hypertension. Antioxid Redox Signal. 2014;20(2):281–94.
44. Rey FE, Cifuentes ME, Kiarash A, Quinn MT, Pagano PJ. Novel competitive inhibitor of NAD(P)H oxidase assembly attenuates vascular O(2)(−) and systolic blood pressure in mice. Circ Res. 2001;89(5):408–14.
45. Martinez-Revelles S, Avendano MS, Garcia-Redondo AB, Alvarez Y, Aguado A, Perez-Giron JV, et al. Reciprocal relationship between reactive oxygen species and cyclooxygenase-2 and vascular dysfunction in hypertension. Antioxid Redox Signal. 2013;18(1):51–65.
46. Schroder K, Zhang M, Benkhoff S, Mieth A, Pliquett R, Kosowski J, et al. Nox4 is a protective reactive oxygen species generating vascular NADPH oxidase. Circ Res. 2012; 110(9):1217–25.

47. Landmesser U, Cai H, Dikalov S, McCann L, Hwang J, Jo H, et al. Role of p47(phox) in vascular oxidative stress and hypertension caused by angiotensin II. Hypertension. 2002;40(4):511–5.
48. Ebrahimian T, Li MW, Lemarie CA, Simeone SM, Pagano PJ, Gaestel M, et al. Mitogen-activated protein kinase-activated protein kinase 2 in angiotensin II-induced inflammation and hypertension: regulation of oxidative stress. Hypertension. 2011;57(2):245–54.
49. Matsui R, Xu S, Maitland KA, Hayes A, Leopold JA, Handy DE, et al. Glucose-6 phosphate dehydrogenase deficiency decreases the vascular response to angiotensin II. Circulation. 2005;112(2):257–63.
50. Cifuentes ME, Rey FE, Carretero OA, Pagano PJ. Upregulation of p67(phox) and gp91(phox) in aortas from angiotensin II-infused mice. Am J Physiol Heart Circ Physiol. 2000; 279(5):H2234–40.
51. Modlinger P, Chabrashvili T, Gill PS, Mendonca M, Harrison DG, Griendling KK, et al. RNA silencing in vivo reveals role of p22phox in rat angiotensin slow pressor response. Hypertension. 2006;47(2):238–44.
52. Fujii A, Nakano D, Katsuragi M, Ohkita M, Takaoka M, Ohno Y, et al. Role of gp91phox-containing NADPH oxidase in the deoxycorticosterone acetate-salt-induced hypertension. Eur J Pharmacol. 2006;552(1-3):131–4.
53. Chrissobolis S, Drummond GR, Faraci FM, Sobey CG. Chronic aldosterone administration causes Nox2-mediated increases in reactive oxygen species production and endothelial dysfunction in the cerebral circulation. J Hypertens. 2014;32(9):1815–21.
54. Viel EC, Benkirane K, Javeshghani D, Touyz RM, Schiffrin EL. Xanthine oxidase and mitochondria contribute to vascular superoxide anion generation in DOCA-salt hypertensive rats. Am J Physiol Heart Circ Physiol. 2008;295(1):H281–8.
55. Beswick RA, Dorrance AM, Leite R, Webb RC. NADH/NADPH oxidase and enhanced superoxide production in the mineralocorticoid hypertensive rat. Hypertension. 2001; 38(5):1107–11.
56. Jimenez R, Lopez-Sepulveda R, Kadmiri M, Romero M, Vera R, Sanchez M, et al. Polyphenols restore endothelial function in DOCA-salt hypertension: role of endothelin-1 and NADPH oxidase. Free Radic Biol Med. 2007;43(3):462–73.
57. Chen QZ, Han WQ, Chen J, Zhu DL, Chen Y, Gao PJ. Anti-stiffness effect of apocynin in deoxycorticosterone acetate-salt hypertensive rats via inhibition of oxidative stress. Hypertens Res. 2013;36(4):306–12.
58. Gomez-Guzman M, Jimenez R, Sanchez M, Zarzuelo MJ, Galindo P, Quintela AM, et al. Epicatechin lowers blood pressure, restores endothelial function, and decreases oxidative stress and endothelin-1 and NADPH oxidase activity in DOCA-salt hypertension. Free Radic Biol Med. 2012;52(1):70–9.
59. Youn JY, Wang T, Blair J, Laude KM, Oak JH, McCann LA, et al. Endothelium-specific sepiapterin reductase deficiency in DOCA-salt hypertension. Am J Physiol Heart Circ Physiol. 2012;302(11):H2243–9.
60. Park JB, Touyz RM, Chen X, Schiffrin EL. Chronic treatment with a superoxide dismutase mimetic prevents vascular remodeling and progression of hypertension in salt-loaded stroke-prone spontaneously hypertensive rats. Am J Hypertens. 2002;15(1 Pt 1):78–84.
61. Briones AM, Tabet F, Callera GE, Montezano AC, Yogi A, He Y, et al. Differential regulation of Nox1, Nox2 and Nox4 in vascular smooth muscle cells from WKY and SHR. J Am Soc Hypertens. 2011;5(3):137–53.
62. Paravicini TM, Chrissobolis S, Drummond GR, Sobey CG. Increased NADPH-oxidase activity and Nox4 expression during chronic hypertension is associated with enhanced cerebral vasodilatation to NADPH in vivo. Stroke. 2004;35(2):584–9.
63. Roque FR, Briones AM, Garcia-Redondo AB, Galan M, Martinez-Revelles S, Avendano MS, et al. Aerobic exercise reduces oxidative stress and improves vascular changes of small mesenteric and coronary arteries in hypertension. Br J Pharmacol. 2013;168(3):686–703.
64. Ren Y, D'Ambrosio MA, Liu R, Pagano PJ, Garvin JL, Carretero OA. Enhanced myogenic response in the afferent arteriole of spontaneously hypertensive rats. Am J Physiol Heart Circ Physiol. 2010;298(6):H1769–75.

65. Zarzuelo MJ, Jimenez R, Galindo P, Sanchez M, Nieto A, Romero M, et al. Antihypertensive effects of peroxisome proliferator-activated receptor-beta activation in spontaneously hypertensive rats. Hypertension. 2011;58(4):733–43.
66. Wind S, Beuerlein K, Armitage ME, Taye A, Kumar AH, Janowitz D, et al. Oxidative stress and endothelial dysfunction in aortas of aged spontaneously hypertensive rats by NOX1/2 is reversed by NADPH oxidase inhibition. Hypertension. 2010;56(3):490–7.
67. Podlutsky A, Ballabh P, Csiszar A. Oxidative stress and endothelial dysfunction in pulmonary arteries of aged rats. Am J Physiol Heart Circ Physiol. 2010;298(2):H346–51.
68. Akasaki T, Ohya Y, Kuroda J, Eto K, Abe I, Sumimoto H, et al. Increased expression of gp91phox homologues of NAD(P)H oxidase in the aortic media during chronic hypertension: involvement of the renin-angiotensin system. Hypertens Res. 2006;29(10):813–20.
69. Zimmerman MC, Lazartigues E, Sharma RV, Davisson RL. Hypertension caused by angiotensin II infusion involves increased superoxide production in the central nervous system. Circ Res. 2004;95(2):210–6.
70. Peterson JR, Burmeister MA, Tian X, Zhou Y, Guruju MR, Stupinski JA, et al. Genetic silencing of Nox2 and Nox4 reveals differential roles of these NADPH oxidase homologues in the vasopressor and dipsogenic effects of brain angiotensin II. Hypertension. 2009;54(5): 1106–14.
71. Lob HE, Schultz D, Marvar PJ, Davisson RL, Harrison DG. Role of the NADPH oxidases in the subfornical organ in angiotensin II-induced hypertension. Hypertension. 2013; 61(2):382–7.
72. Nozoe M, Hirooka Y, Koga Y, Sagara Y, Kishi T, Engelhardt JF, et al. Inhibition of Rac1-derived reactive oxygen species in nucleus tractus solitarius decreases blood pressure and heart rate in stroke-prone spontaneously hypertensive rats. Hypertension. 2007;50(1):62–8.
73. Faraci FM, Didion SP. Vascular protection: superoxide dismutase isoforms in the vessel wall. Arterioscler Thromb Vasc Biol. 2004;24:1367–73.
74. Fukai T, Ushio-Fukai M. Superoxide dismutases: role in redox signaling, vascular function, and diseases. Antioxid Redox Signal. 2011;15(6):1583–606.
75. Didion SP, Ryan MJ, Didion LA, Fegan PE, Sigmund CD, Faraci FM. Increased superoxide and vascular dysfunction in CuZnSOD-deficient mice. Circ Res. 2002;91(10):938–44.
76. Baumbach GL, Didion SP, Faraci FM. Hypertrophy of cerebral arterioles in mice deficient in expression of the gene for CuZn superoxide dismutase. Stroke. 2006;37(7):1850–5.
77. Didion SP, Kinzenbaw DA, Schrader LI, Faraci FM. Heterozygous CuZn superoxide dismutase deficiency produces a vascular phenotype with aging. Hypertension. 2006;48(6):1072–9.
78. Wang HD, Johns DG, Xu S, Cohen RA. Role of superoxide anion in regulating pressor and vascular hypertrophic response to angiotensin II. Am J Physiol Heart Circ Physiol. 2002;282(5):H1697–702.
79. Li Y, Huang TT, Carlson EJ, Melov S, Ursell PC, Olson JL, et al. Dilated cardiomyopathy and neonatal lethality in mutant mice lacking manganese superoxide dismutase. Nat Genet. 1995;11(4):376–81.
80. Suzuki K, Tatsumi H, Satoh S, Senda T, Nakata T, Fujii J, et al. Manganese-superoxide dismutase in endothelial cells: localization and mechanism of induction. Am J Physiol Heart Circ Physiol. 1993;265(4 Pt 2):H1173–8.
81. Uddin M, Yang H, Shi M, Polley-Mandal M, Guo Z. Elevation of oxidative stress in the aorta of genetically hypertensive mice. Mech Ageing Dev. 2003;124:811–7.
82. Ulker S, McMaster D, McKeown PP, Bayraktutan U. Impaired activities of antioxidant enzymes elicit endothelial dysfunction in spontaneous hypertensive rats despite enhanced vascular nitric oxide generation. Cardiovasc Res. 2003;59:488–500.
83. Rodriguez-Iturbe B, Sepassi L, Quiroz Y, Ni Z, Vaziri ND. Association of mitochondrial SOD deficiency with salt-sensitive hypertension and accelerated renal senescence. J Appl Physiol. 2007;102(1):255–60.
84. Chrissobolis S, Faraci FM. Sex differences in protection against angiotensin ii-induced endothelial dysfunction by manganese superoxide dismutase in the cerebral circulation. Hypertension. 2010;55(4):905–10.

85. Kimura S, Zhang GX, Nishiyama A, Shokoji T, Yao L, Fan YY, et al. Mitochondria-derived reactive oxygen species and vascular MAP kinases: comparison of angiotensin II and diazoxide. Hypertension. 2005;45(3):438–44.
86. Nazarewicz RR, Dikalova AE, Bikineyeva A, Dikalov SI. Nox2 as a potential target of mitochondrial superoxide and its role in endothelial oxidative stress. Am J Physiol Heart Circ Physiol. 2013;305(8):H1131–40.
87. Dikalova AE, Bikineyeva AT, Budzyn K, Nazarewicz RR, McCann L, Lewis W, et al. Therapeutic targeting of mitochondrial superoxide in hypertension. Circ Res. 2010;107(1): 106–16.
88. Wenzel P, Schuhmacher S, Kienhofer J, Muller J, Hortmann M, Oelze M, et al. Manganese superoxide dismutase and aldehyde dehydrogenase deficiency increase mitochondrial oxidative stress and aggravate age-dependent vascular dysfunction. Cardiovasc Res. 2008; 80(2):280–9.
89. Roos CM, Hagler M, Zhang B, Oehler EA, Arghami A, Miller JD. Transcriptional and phenotypic changes in aorta and aortic valve with aging and MnSOD deficiency in mice. Am J Physiol Heart Circ Physiol. 2013;305(10):H1428–39.
90. Gongora MC, Qin Z, Laude K, Kim HW, McCann L, Folz JR, et al. Role of extracellular superoxide dismutase in hypertension. Hypertension. 2006;48(3):473–81.
91. Jung O, Marklund SL, Geiger H, Pedrazzini T, Busse R, Brandes RP. Extracellular superoxide dismutase is a major determinant of nitric oxide bioavailability: in vivo and ex vivo evidence from ecSOD-deficient mice. Circ Res. 2003;93(7):622–9.
92. Qin Z, Gongora MC, Ozumi K, Itoh S, Akram K, Ushio-Fukai M, et al. Role of Menkes ATPase in angiotensin II-induced hypertension: a key modulator for extracellular superoxide dismutase function. Hypertension. 2008;52(5):945–51.
93. Ozumi K, Sudhahar V, Kim HW, Chen GF, Kohno T, Finney L, et al. Role of copper transport protein antioxidant 1 in angiotensin II-induced hypertension: a key regulator of extracellular superoxide dismutase. Hypertension. 2012;60(2):476–86.
94. Lob HE, Vinh A, Li L, Blinder Y, Offermanns S, Harrison DG. Role of vascular extracellular superoxide dismutase in hypertension. Hypertension. 2011;58(2):232–9.
95. Lubos E, Kelly NJ, Oldebeken SR, Leopold JA, Zhang YY, Loscalzo J, et al. Glutathione peroxidase-1 deficiency augments proinflammatory cytokine-induced redox signaling and human endothelial cell activation. J Biol Chem. 2011;286(41):35407–17.
96. Hoen PA, Van der Lans CA, Van Eck M, Bijsterbosch MK, Van Berkel TJ, Twisk J. Aorta of ApoE-deficient mice responds to atherogenic stimuli by a prelesional increase and subsequent decrease in the expression of antioxidant enzymes. Circ Res. 2003;93(3):262–9.
97. Ardanaz N, Yang XP, Cifuentes ME, Haurani MJ, Jackson KW, Liao TD, et al. Lack of glutathione peroxidase 1 accelerates cardiac-specific hypertrophy and dysfunction in angiotensin II hypertension. Hypertension. 2010;55(1):116–23.
98. Forgione MA, Cap A, Liao R, Moldovan NI, Eberhardt RT, Lim CC, et al. Heterozygous cellular glutathione peroxidase deficiency in the mouse: abnormalities in vascular and cardiac function and structure. Circulation. 2002;106(9):1154–8.
99. Chrissobolis S, Didion SP, Kinzenbaw DA, Schrader LI, Dayal S, Lentz SR, et al. Glutathione peroxidase-1 plays a major role in protecting against angiotensin II-induced vascular dysfunction. Hypertension. 2008;51:872–7.
100. Oelze M, Kroller-Schon S, Steven S, Lubos E, Doppler C, Hausding M, et al. Glutathione peroxidase-1 deficiency potentiates dysregulatory modifications of endothelial nitric oxide synthase and vascular dysfunction in aging. Hypertension. 2014;63(2):390–6.
101. Ruseva B, Atanasova M, Georgieva M, Shumkov N, Laleva P. Effects of selenium on the vessel walls and anti-elastin antibodies in spontaneously hypertensive rats. Exp Biol Med. 2012;237(2):160–6.
102. Guzik TJ, Hoch NE, Brown KA, McCann LA, Rahman A, Dikalov S, et al. Role of the T cell in the genesis of angiotensin II induced hypertension and vascular dysfunction. J Exp Med. 2007;204(10):2449–60.
103. Ji H, Zheng W, Li X, Liu J, Wu X, Zhang MA, et al. Sex-specific T-cell regulation of angiotensin II-dependent hypertension. Hypertension. 2014;64(3):573–82.

104. Hausding M, Jurk K, Daub S, Kroller-Schon S, Stein J, Schwenk M, et al. CD40L contributes to angiotensin II-induced pro-thrombotic state, vascular inflammation, oxidative stress and endothelial dysfunction. Basic Res Cardiol. 2013;108(6):386.
105. Wenzel P, Knorr M, Kossmann S, Stratmann J, Hausding M, Schuhmacher S, et al. Lysozyme M-positive monocytes mediate angiotensin II-induced arterial hypertension and vascular dysfunction. Circulation. 2011;124(12):1370–81.
106. Montezano AC, Touyz RM. Oxidative stress, Noxs, and hypertension: experimental evidence and clinical controversies. Ann Med. 2012;44 Suppl 1:S2–16.
107. Guzik TJ, Chen W, Gongora MC, Guzik B, Lob HE, Mangalat D, et al. Calcium-dependent NOX5 nicotinamide adenine dinucleotide phosphate oxidase contributes to vascular oxidative stress in human coronary artery disease. J Am Coll Cardiol. 2008;52(22):1803–9.
108. Touyz RM, Schiffrin EL. Increased generation of superoxide by angiotensin II in smooth muscle cells from resistance arteries of hypertensive patients: role of phospholipase D-dependent NAD(P)H oxidase-sensitive pathways. J Hypertens. 2001;19(7):1245–54.
109. San Jose G, Moreno MU, Olivan S, Beloqui O, Fortuno A, Diez J, et al. Functional effect of the p22phox -930A/G polymorphism on p22phox expression and NADPH oxidase activity in hypertension. Hypertension. 2004;44(2):163–9.
110. Schreiber R, Ferreira-Sae MC, Ronchi JA, Pio-Magalhaes JA, Cipolli JA, Matos-Souza JR, et al. The C242T polymorphism of the p22-phox gene (CYBA) is associated with higher left ventricular mass in Brazilian hypertensive patients. BMC Med Genet. 2011;12:114.
111. Moreno MU, San Jose G, Fortuno A, Beloqui O, Diez J, Zalba G. The C242T CYBA polymorphism of NADPH oxidase is associated with essential hypertension. J Hypertens. 2006;24(7):1299–306.
112. Moreno MU, San Jose G, Fortuno A, Beloqui O, Redon J, Chaves FJ, et al. A novel CYBA variant, the -675A/T polymorphism, is associated with essential hypertension. J Hypertens. 2007;25(8):1620–6.

Chapter 5
Oxidative Stress and Central Regulation of Blood Pressure

Yoshitaka Hirooka and Kenji Sunagawa

Introduction

There is a growing body of evidence indicating that reactive oxygen species (ROS) such as superoxide anions and hydroxyl radicals play an important role in the pathogenesis of hypertension [4, 9, 11–13, 28] and that ROS production is increased in many types of hypertensive models [13, 28]. Among the target organs of hypertensive diseases, the brain is most affected by ageing and oxidative stress. We and other investigators have demonstrated that ROS in the brain are responsible for increased sympathetic activation in animal models of hypertension and heart failure [2, 3, 5, 8, 9, 11–13]. The central nervous system contains several important nuclei for controlling sympathetic activity [6, 7], with the rostral ventrolateral medulla (RVLM) in the brainstem being the vasomotor center that determines basal sympathetic nerve activity, and the functional integrity of the RVLM is essential for the maintenance of basal vasomotor tone. Other nuclei such as the nucleus tractus solitarius (NTS), paraventricular nucleus of the hypothalamus (PVN), and subfornical organ are also important for autonomic blood pressure control [5, 6, 12, 13]. In this chapter, we discuss a series of studies regarding the role of ROS in the central nervous system control of blood pressure and hypertension.

Y. Hirooka, M.D., Ph.D. (✉)
Department of Advanced Cardiovascular Regulation and Therapeutics,
Center for Disruptive Cardiovascular Medicine, Kyushu University,
3-1-1 Maidashi, Higashi-ku, Fukuoka 812-8582, Japan
e-mail: hyoshi@cardiol.med.kyushu-u.ac.jp

K. Sunagawa
Department of Therapeutic Regulation of Cardiovascular Homeostasis, Center for Disruptive Cardiovascular Medicine, Kyushu University, Fukuoka 812-85882, Japan

M. Rodriguez-Porcel et al. (eds.), *Studies on Atherosclerosis*,
Oxidative Stress in Applied Basic Research and Clinical Practice,
DOI 10.1007/978-1-4899-7693-2_5

Increased ROS in the RVLM in Stroke-Prone Spontaneously Hypertensive Rats

We compared ROS levels, evaluated by measuring thiobarbituric acid-reactive substances (TBARS), in the RVLM of stroke-prone spontaneously hypertensive rats (SHRSP) with those in normotensive Wistar-Kyoto (WKY) rats [18, 35]. TBARS are end products of lipid peroxidation and an indirect marker of oxidative stress. ROS levels were also measured by electron spin resonance spectroscopy and we found that the electron spin resonance signal decay rate in the RVLM of SHRSP was significantly increased compared with that in WKY rats. By contrast, superoxide dismutase (SOD) activity in the RVLM was decreased in SHRSP compared with WKY rats. Functionally, bilateral injection of Tempol into the RVLM of SHRSP decreased blood pressure, but the same response was not observed in WKY rats. Furthermore, overexpression of manganese SOD (MnSOD) in the RVLM of SHRSP decreased blood pressure and sympathetic activity. These findings indicate that ROS in the RVLM are increased in SHRSP and contribute to the neural mechanisms of hypertension in SHRSP. These effects are mediated by an increase in γ-amino butyric acid (GABA)ergic inhibitory inputs to the RVLM neurons in SHRSP [36]. The increase in ROS production in the RVLM has been shown in SHR from other laboratories [2, 3]. It should be noted that oxidative stress in the hypothalamus and subfornical organ plays a significant role in the pathogenesis of hypertension [4, 5]. These areas influence not only sympathetic activation, but drinking behavior and vasopressin release sending the signal to the supraoptic nucleus and PVN [4, 6]. This is also apparent in angiotensin II-induced hypertension [4].

Sources of ROS Production in the Brain

There are several sources of ROS production including NAD(P)H oxidase, mitochondria, xanthine oxidase, and uncoupled nitric oxidase synthase. Among them, we found that the Rac1/NAD(P)H oxidase pathway is activated in the brainstem of SHRSP thereby eliciting ROS production [11, 33]. Rac1 is a small G protein that is an important signaling molecule involved in integrating intracellular transduction pathways toward NAD(P)H oxidase activation. Furthermore, Rac1 requires lipid modifications to migrate from the cytosol to the plasma membrane, which is a necessary step for activating the ROS-generating NAD(P)H oxidase enzyme system. We assessed Rac1-GTP levels as an index of Rac1 activation using a glutathione S-transferase-PAK pull-down assay and found that Rac1 activity in the NTS and RVLM of SHRSP was increased compared with those of WKY rats. Consistent with increased Rac1 activation, NAD(P)H oxidase-dependent superoxide production was also increased in the brainstem of SHRSP compared with those of WKY rats. Furthermore, gene transfer of adenovirus vectors encoding dominant-negative Rac1 into the NTS or RVLM suppressed both Rac1 and NAD(P)H oxidase activity and decreased blood pressure, heart rate, and urinary norepinephrine excretion in SHRSP [11, 34]. Chan et al. demonstrated that NADPH oxidase subunits including gp91phox and p22phox are upregulated in the RVLM of SHR [3]. In addition, we observed that Cu/Zn-SOD expression and activity in the NTS was decreased in the NTS of SHRSP compared with WKY rats. Functionally, gene transfer of Cu/Zn-SOD into the NTS also decreased in SHRSP, but not in WKY rats. Oxidative stress evaluated by TBARS in the brainstem was increased in SHRSP and gene transfer of dominant negative Rac1 or Cu/Zn-SOD reduced those levels. Taken together, these findings indicate that activation of Rac1/NAD(P)H oxidase in the brainstem of SHRSP is one of the sources of ROS production thereby increasing blood pressure via the sympathetic nervous system. It has been previously shown that angiotensin II type 1 receptors (AT1R) are upstream of NAD(P)H oxidase, suggesting that AT1R activation leads to ROS production. In fact, we found that AT1R activation also increases ROS production from the mitochondria, suggesting that mitochondria are another important source of ROS in the RVLM [34].

Other sources of ROS production that have been demonstrated in hypertension include xanthine oxidase and nitric oxide synthase (NOS) uncoupling. In fact, we demonstrated that overexpression of inducible NOS (iNOS) in the RVLM elicits hypertension and sympathoexcitation via an increase in ROS production [16]. Increased ROS production was confirmed by lipid peroxidation in the RVLM by TBARS and dihydroethidium staining. Furthermore, we found that iNOS expression level in the RVLM is increased in SHR and microinjection of iNOS inhibitors into the RVLM reduced blood pressure in SHR [17].

We also investigated the downstream pathway of AT1R stimulation in the RVLM of SHRSP and found that the Ras, p38 mitogen-activated protein kinase, extracellular signal-regulated kinase, apoptotic proteins Bax and Bad, and caspase-3 in the RVLM are activated in SHRSP [21]. Chronic intracerebroventricular infusion of the Ras inhibitor or the caspase-3 inhibitor reduced blood pressure in SHRSP, but not in WKY rats, suggesting these pathways are involved in activation of sympathetic activity via AT1R stimulation. As a downstream signaling cascade, redox-sensitive upregulation of transcriptional factors, such as activator protein-1, c-jun is observed in the RVLM of hypertensive rats [2, 3].

We elucidated whether ROS in the RVLM modulate synaptic transmission via excitatory and inhibitory amino acids and influence the excitatory inputs to the RVLM from the PVN in SHR [30]. We found that ROS in the RVLM enhanced glutamatergic excitatory inputs and attenuated the GABAergic inhibitory inputs to the RVLM neurons in SHR. Furthermore, ROS in the RVLM enhanced the pressor and sympathoexcitatory response induced by activation of the PVN neurons. Our findings indicate that increased ROS in the RVLM of SHR contribute to hypertension by altering synaptic transmission in the RVLM through sympathoexcitation evoked by enhancing glutamatergic inputs to the RVLM from the PVN and attenuating the GABA-mediated sympathoinhibition in the RVLM. Then, we investigated the relative contribution of oxidative stress in the PVN and RVLM of SHR in blood pressure regulation. For this purpose, we transfected adenovirus vectors encoding the MnSOD gene (AdMnSOD) or β-galactosidase gene (AdLacZ) bilaterally into the RVLM or PVN [31]. Blood pressure and heart rate of AdMnSOD–RVLM-transfected SHR were decreased compared with AdLacZ–RVLM-transfected SHR. In contrast, blood pressure of AdMnSOD–PVN-transfected SHR was not decreased compared with AdLacZ–PVN-transfected SHR, but heart rate was decreased compared with AdLacZ–PVN-transfected SHR. MnSOD transfection into both the RVLM and PVN of SHR decreased blood pressure and elicited a profound decrease in heart rate. Therefore, oxidative stress in the PVN and RVLM plays a different role for cardiovascular regulation in SHR via the autonomic nervous system. It is also suggested that endoplasmic reticulum stress plays an important role in the generation of the ROS particularly in angiotensin II-induced hypertension [4].

High Salt Intake and ROS in the RVLM of Hypertensive Rats

It is known that high salt intake is an important environmental factor for human hypertension and that genetic factors also might be involved in salt sensitivity. Therefore, many studies have been performed using experimental hypertensive models. We demonstrated that high salt intake enhances blood pressure increase during development of hypertension via oxidative stress in the RVLM of SHR [25]. In SHR, oxidative stress in the RVLM is increased, thereby this mechanism is involved in hypertension of SHR. In addition, further increases in oxidative stress in the RVLM augment the development of hypertension of SHR. Changes were correlated to the AT1R expression levels in the RVLM suggesting that ROS production is increased associated with activation of AT1R and NAD(P)H oxidase. This concept is supported by other investigators and summarized well [5]. The PVN may also play an important role in salt-sensitive hypertension and mineralocorticoid receptors in the brain could be activated [5].

Recently, we found that moxonidine-induced central sympathoinhibition attenuated brain oxidative stress, prevented cardiac dysfunction and remodeling, and improved the prognosis in rats with hypertensive heart failure [14]. Dahl salt-sensitive rats fed a high salt diet from 7 weeks of age served as the hypertensive heart failure model in that study. We suggest that central sympathoinhibition can be effective for the treatment of hypertensive heart failure.

ROS in the RVLM of Obesity-Induced Hypertensive Rats

Metabolic syndrome (MetS) is characterized by the presence of central obesity, impaired fasting glucose, dyslipidemia, and obesity-induced hypertension. It is suggested that activation of the sympathetic nervous system is the major cause of obesity-induced hypertension [37]. We found that oxidative stress in the RVLM and sympathetic activity are increased in obesity-prone rats manifesting a MetS profile.

Furthermore, orally administered telmisartan, an ARB, inhibited sympathetic activity through antioxidant effects via AT1R and oxidative stress in the RVLM of obesity-prone rats [27]. Furthermore, we found that calorie restriction inhibits sympathetic activity through antioxidant effects in the RVLM, although it is not clear whether activation of AT1R is attenuated by calorie restriction [22]. We suggest that AT1R and oxidative stress in the RVLM might be novel therapeutic targets for obesity-induced hypertension through sympathoinhibition, and an ARB more accessible for AT1R in the RVLM might be preferable for obesity-induced hypertension. It is also suggested that oxidative stress also plays an important role in obesity-induced hypertension probably because of inflammatory pathways [5].

Possibility of Currently Used Cardiovascular Drugs Acting on the Brain Thereby Reducing Central Sympathetic Outflow

Some currently used antihypertensive drugs have antioxidant effects in the brain, in particular, the RVLM thereby reducing enhanced central sympathetic outflow which might explain why such drugs do not elicit reflex-mediated sympathoexcitation. As previously described, because activation of AT1R in the brain is the main source of ROS production, AT1R blockers (ARBs) are candidates for capability of antioxidant actions in the RVLM. We demonstrated that telmisartan, olmesartan, and irbesartan might have antioxidant effects acting on the RVLM thereby reducing sympathetic activity in hypertensive rats [1, 11, 23, 32]. However, it is not known whether ARBs are able to cross the blood–brain barrier thereby directly acting on the RVLM. Furthermore, pharmacokinetics and lipophilicity of these compounds, and the transport system between the brain and blood, may be some of the factors that may limit their action on the RVLM. Despite the potential limitations of these compounds, we recently found that treatment with telmisartan reduced mortality and left ventricular hypertrophy with sympathoinhibition in angiotensin II-infused and high salt intake SHRSP [24].

In addition, it has been known that the production of ROS in the brain is somewhat linked to L-type calcium channels. Therefore, the long-acting and lipophilic calcium channel blockers are possible candidates for antioxidant effects in the brain. At least, we found that Azelnidipine might have such effects from experimental studies in hypertensive rats [26]. Furthermore, Azelnidipine is suggested to decrease heart rate probably because of sympathoinhibitory action in patients with hypertension. The demerit of calcium channel blockers is reflex-mediated sympathetic activation due to strong vasodilatory action. This is particularly true in the case of short-acting calcium channel blockers. Amlodipine, but not nicardipine, has some effects in experimental hypertensive rats [10]. However, in human hypertensive patients, amlodipine slightly increases sympathetic activity, even its long half-life.

We also have found that atorvastatin reduces oxidative stress in the RVLM of SHRSP thereby leading to reduction of blood pressure via inhibition of sympathetic activity [19]. Furthermore, statins suggested increases in NO production, an effect that might be related to an antioxidant effect of statins. In fact, we also found that atorvastatin improves the impaired baroreflex sensitivity measured by the spontaneous sequence method via the antioxidant effect in the RVLM [20].

Conclusions

In this chapter, we show the series of our studies regarding central control of blood pressure with respect to oxidative stress in the brain, particularly the RVLM (Fig. 5.1). We suggest that AT1R in the RVLM are a therapeutic target for hypertension as well as heart failure in which the enhanced central sympathetic drive is responsible for worsening the disease process. Recently we extended this possibility to atrial fibrillation [29]. We demonstrated that brain AT1R blockade suppresses atrial fibrillation inducibility and maintenance independent of depressor response in hypertensive rats [29]. Activation of the AT1R continued on NAD(P)H oxidase is the major source of ROS in the RVLM. Recently, we reported that AT1R expression

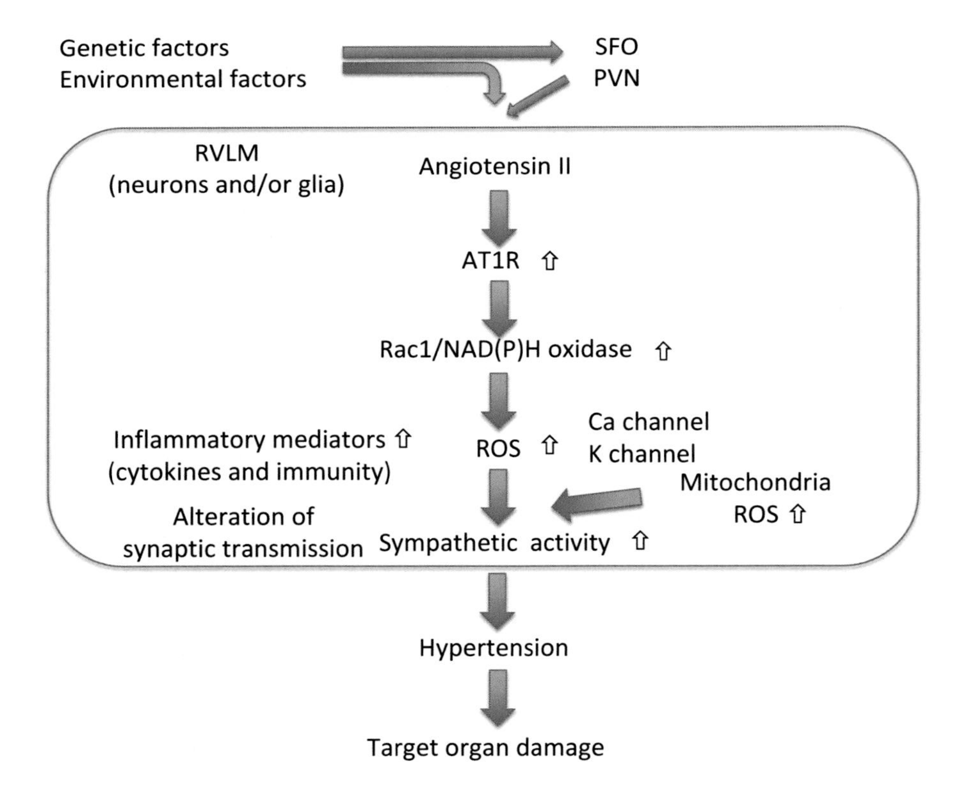

Fig. 5.1 A scheme demonstrating that AT1R activation elicits ROS production thereby causing hypertension via sympathetic activation in the RVLM

in astrocytes is upregulated in mice with myocardial-infarction–induced heart failure [14]. Interestingly, we found that targeting AT1R in astrocytes using the Cre-LoxP system, and deletion of AT1R in astrocytes improved the remodeling process and survival after myocardial infarction [15]. Further investigations are warranted for understanding the precise mechanisms involved and application of this concept in humans. Recent studies are expanding the field of the inflammatory process and immunity related to ROS production [4, 13]. Finally, it should be noted that ROS in the kidney and vasculature are also critically involved in the pathogenesis of hypertension [4, 28]. We further need to explore the clinical relevance of increased ROS in human hypertension from accumulating evidence of experimental studies.

Acknowledgments The series of studies was supported by Grants-in Aid for Scientific Research from the Japan Society for the Promotion of Science (B193290231, B24390198). The Department of Advanced Cardiovascular Regulation and Therapeutics, Kyushu University is supported by Actelion Pharmaceuticals Ltd (Y.H.).

References

1. Araki S, Hirooka Y, Kishi T, Yasukawa K, Utsumi H, Sunagawa K. Olmesartan reduces oxidative stress in the brain of stroke-prone spontaneously hypertensive rats assessed by an in vivo ESR method. Hypertens Res. 2009;32:1091–6.
2. Chan SHH, Chan JYH. Brain stem oxidative stress and its associated signaling in the regulation of sympathetic vasomotor tone. J Appl Physiol. 2012;113:1921–8.
3. Chan SHH, Chan JYH. Brain stem NOS and ROS in neural mechanisms of hypertension. Antioxid Redox Signal. 2014;20:146–63.
4. Francis J, Davisson RL. Emerging concepts in hypertension. Antioxid Redox Signal. 2014;20:69–73.
5. Fujita M, Fujita T. The role of CNS in salt-sensitive hypertension. Curr Hypertens Rep. 2013;15:390–4.
6. Gabor A, Leenen FHH. Central neuromodulatory pathways regulating sympathetic activity in hypertension. J Appl Physiol. 2012;113:1294–303.
7. Hirooka Y. Localized gene transfer and its application for the study of central cardiovascular control. Auton Neurosci. 2006;126/127:120–9.
8. Hirooka Y. Role of reactive oxygen species in brainstem in neural mechanisms of hypertension. Auton Neurosci. 2008;142:20–4.
9. Hirooka Y. Oxidative stress in the cardiovascular center has a pivotal role in the sympathetic activation in hypertension. Hypertens Res. 2011;34:407–12.
10. Hirooka Y, Kimura Y, Nozoe M, Sagara Y, Ito K, Sunagawa K. Amlodipine-induced reduction of oxidative stress in the brain is associated with sympatho-inhibitory effects in stroke-prone spontaneously hypertensive rats. Hypertens Res. 2006;29:49–56.
11. Hirooka Y, Sagara Y, Kishi T, Sunagawa K. Oxidative stress and central cardiovascular regulation-pathogenesis of hypertension and therapeutic aspects. Circ J. 2010;74:827–35.
12. Hirooka Y, Kishi T, Sakai K, Takeshita A, Sunagawa K. Imbalance of central nitric oxide and reactive oxygen species in the regulation of sympathetic activity and neural mechanisms of hypertension. Am J Physiol Regul Integr Comp Physiol. 2011;300:R818–26.
13. Hirooka Y, Kishi T, Ito K, Sunagawa K. Potential clinical application of recently discovered brain mechanisms involved in hypertension. Hypertension. 2013;62:995–1002.
14. Honda N, Hirooka Y, Ito K, Matsukawa R, Shinohara K, Kishi T, et al. Moxonidine-induced central sympathoinhibition improves prognosis in rats with hypertensive heart failure. J Hypertens. 2013;31:2300–8.

15. Isegawa K, Hirooka Y, Katsuki M, Kishi T, Sunagawa K. Angiotensin type 1 receptor expression in astrocytes is upregulated leading to increased mortality in mice with myocardial infarction-induced heart failure. Am J Physiol Heart Circ Physiol. 2014;307:H1448–55.
16. Kimura Y, Hirooka Y, Sagara Y, Ito K, Kishi T, Shimokawa H, et al. Ocerexpression of inducible nitric oxide synthase in rostral ventrolateral medulla causes hypertension and sympathoexcitation via an increase in oxidative stress. Circ Res. 2005;96:252–60.
17. Kimura Y, Hirooka Y, Kishi T, Ito K, Sagara Y, Sunagawa K. Role of inducible synthase in rostral ventrolateral medulla in blood pressure regulation in spontaneously hypertensive rats. Clin Exp Hypertens. 2009;31:281–6.
18. Kishi T, Hirooka Y, Kimura Y, Ito K, Shimokawa H, Takeshita A. Increased reactive oxygen species in rostral ventrolaterla medulla contribute to neural mechanisms of hypertension in stroke-prone spontaneously hypertensive rats. Circulation. 2004;109:2357–62.
19. Kishi T, Hirooka Y, Shimokawa H, Takeshita A, Sunagawa K. Atorvastatin reduces oxidative stress in the rostral ventrolateral medulla of stroke-prone spontaneously hypertensive rats. Clin Exp Hypertens. 2008;30:3–11.
20. Kishi T, Hirooka Y, Konno S, Sunagawa K. Atorvastatin improves the impaired baroreflex sensitivity via anti-oxidant effect in the rostral ventrolateral medulla of SHRSP. Clin Exp Hypertens. 2009;31:698–704.
21. Kishi T, Hirooka Y, Konno S, Ogawa K, Sunagawa K. Angiotensin II type 1 receptor-activated caspase-3 through ras/mitogen-activated protein kinase/extracellular signal-regulated kinase in the rostral ventrolateral medulla is involved in sympathoexcitation in stroke-prone spontaneously hypertensive rats. Hypertension. 2010;55:291–7.
22. Kishi T, Hirooka Y, Ogawa K, Konno S, Sunagawa K. Calorie restriction inhibits sympathetic nerve activity via anti-oxidant effect in the rostral ventrolateral medulla of obesity-induced hypertensive rats. Clin Exp Hypertens. 2011;33:240–5.
23. Kishi T, Hirooka Y, Sunagawa K. Sympathoinhibition caused by orally administered telmisartan through inhibition of the AT1 receptor in the rostral ventrolateral medulla of hypertensive rats. Hypertens Res. 2012;35:940–6.
24. Kishi T, Hirooka Y, Sunagwa K. Telmisartan reduces mortality and left ventricular hypertrophy with sympathoinhibition in rats with hypertension and heart failure. Am J Hypertens. 2014;27:260–7.
25. Koga Y, Hirooka Y, Araki S, Nozoe M, Kishi T, Sunagawa K. High salt intake enhances blood pressure increase during development of hypertension via oxidative stress in rostral ventrolateral medulla of spontaneously hypertensive rats. Hypertens Res. 2008;31:2075–83.
26. Konno S, Hirooka Y, Araki S, Koga Y, Kishi T, Sunagawa K. Azelnidipine decreases sympathetic nerve activity via antioxidant effect in the rostral ventrolateral medulla of stroke-prone spontaneously hypertensive rats. J Cardiovasc Pharamacol. 2008;52:555–60.
27. Konno S, Hirooka Y, Kishi T, Sunagawa K. Sympathoinhibitory effects of telmisartan through the reduction of oxidative stress in the rostral ventrolateral medulla of obesity-induced hypertensive rats. J Hypertens. 2012;30:1992–9.
28. Montezano AC, Touyz RM. Reactive oxygen species, vascular Noxs, and hypertension: focus on translational and clinical research. Antioxid Redox Signal. 2014;20:164–82.
29. Nagayama T, Hirooka Y, Kishi T, Mukai Y, Inoue S, Takase S, et al. Blockade of brain angiotensin II type 1 receptor inhibits the development of atrial fibrillation in hypertensive rats. Am J Hypertens. 2015;28:444–51.
30. Nishihara M, Hirooka Y, Matsukawa R, Kishi T, Sunagawa K. Oxidative stress in the rostral ventrolateral medulla modulates excitatory and inhibitory inputs in spontaneously hypertensive rats. J Hypertens. 2012;30:97–106.
31. Nishihara M, Hirooka Y, Kishi T, Sunagawa K. Different role of oxidative stress in paraventricular nucleus and rostral ventrolateral medulla in cardiovascular regulation in awake spontaneously hypertensive rats. J Hypertens. 2012;30:1758–65.
32. Nishihara M, Hirooka Y, Sunagwa K. Combining irebesartan and trichlormethiazide enhances blood pressure reduction via inhibition of sympathetic activity without adverse effects on metabolism in hypertensive rats with metabolic syndrome. Clin Exp Hypertens. 2015;37: 33–8.

33. Nozoe M, Hirooka Y, Koga Y, Sagara Y, Kishi T, Engelhardt JF, et al. Inhibition of rac1-derived reactive oxygen species in nucleus tractus solitarius decreases blood pressure and heart arte in stroke-prone spontaneously hypertensive rats. Hypertension. 2007;50:62–8.
34. Nozoe M, Hirooka Y, Koga Y, Araki S, Konno S, Kishi T, et al. Mitochndria-derived reactive oxygen species mediate sympathoexcitation induced by angiotensin II in the rostral ventrolateral medulla. J Hypertens. 2008;26:2176–84.
35. Ogawa K, Hirooka Y, Shinohara K, Kishi T, Sunagawa K. Inhibition of oxidative stress in rostral ventrolateral medulla improves impaired baroreflex sensitivity in stroke-prone spontaneously hypertensive rats. Int Heart J. 2012;53:193–8.
36. Shinohara K, Hirooka Y, Kishi T, Sunagawa K. Reduction of nitric oxide-mediated γ-amino butyric acid release in rostral ventrolateral medulla is involved in superoxide-induced sympathoexcitation of hypertensive rats. Circ J. 2012;76:2814–21.
37. Sorota S. The sympathetic nervous system as a target for the treatment of hypertension and cardiometabolic diseases. J Cardiovasc Pharamacol. 2014;63:466–76.

Chapter 6
Reactive Oxygen Species and the Regulation of Cerebral Vascular Tone

T. Michael De Silva and Frank M. Faraci

Introduction

The presence of large and/or small vessel disease in the cerebral circulation has a major impact on the brain, greatly increasing the risk for stroke, dementia, and varied neurological diseases. In relation to vascular biology in general and studies of noncerebral blood vessels, a major effort has focused on defining the importance of oxidative stress and oxidant-dependent processes in vascular disease. While lagging behind this broader effort, similar attention has been directed at the cerebral circulation.

Oxidative stress is generally thought to reflect a shift in conditions that favors the generation of reactive oxygen species (ROS) over scavenging of the same molecules by antioxidant defense mechanisms [1]. Oxidative stress can also arise as the result of reduced capacity of antioxidant defense mechanisms. In relation to these mechanisms, cerebral blood vessels have the capacity to generate relatively high levels of superoxide anion, and effects of ROS are prominent in this vascular bed. In some cases, effects of ROS and oxidative stress are greater in the cerebrovasculature compared to other vascular beds [2–4].

In this chapter, we summarize concepts related to the effects of ROS in the cerebral circulation. ROS are known to affect diverse elements of vessel biology including vascular structure [5–7], aneurysm formation and rupture [8], thrombosis [9], and integrity of the blood–brain barrier [10]. In this summary, we will focus on the impact of ROS and oxidative stress in the regulation of cerebrovascular tone.

T.M. De Silva • F.M. Faraci, Ph.D. (✉)
Departments of Internal Medicine and Pharmacology, Francois M. Abboud Cardiovascular Center, Carver College of Medicine, University of Iowa, Iowa City, IA 52242-1081, USA

Iowa City Veterans Affairs Healthcare System, Iowa City, IA 52242-1081, USA
e-mail: frank-faraci@uiowa.edu

M. Rodriguez-Porcel et al. (eds.), *Studies on Atherosclerosis*,
Oxidative Stress in Applied Basic Research and Clinical Practice,
DOI 10.1007/978-1-4899-7693-2_6

Which ROS Are Important?

Oxygen-derived free radicals are a subgroup of ROS containing one or more unpaired electrons and include molecules such as superoxide anion and hydroxyl radical. Although not a free radical, hydrogen peroxide is also included within the larger family of ROS. Interactions between ROS with other molecules are common and can be complex. Superoxide, formed from molecular oxygen in mitochondria and by diverse enzymatic and nonenzymatic sources, is the precursor to many other ROS (Fig. 6.1). Because these interactions are so common, it can be difficult to demonstrate definitively whether specific ROS act in isolation. The molecular properties and functional impact of new ROS species can be fundamentally different from those of the parent molecule (Fig. 6.1).

In addition to relationships with other ROS, superoxide can react with other species resulting in formation of new molecules, some of which are known to be important in relation to vascular disease and regulation of vascular tone (Fig. 6.1). For example, superoxide and nitric oxide (NO) react so efficiently (through an essentially

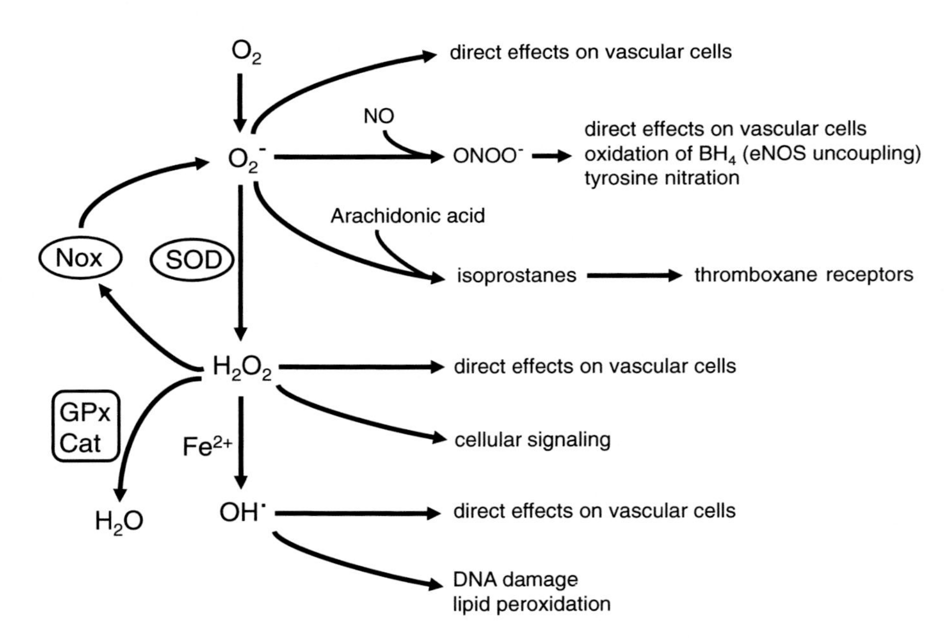

Fig. 6.1 Illustration of the interrelationships of select reactive oxygen species (ROS) that affect vascular tone. Superoxide (O_2^-) is produced from molecular oxygen (O_2) by a variety of sources, including NADPH oxidase (Nox). Superoxide can exert direct effects on vascular cells but can also be converted by superoxide dismutases (SOD) to hydrogen peroxide (H_2O_2). H_2O_2 produces effects of its own or can be degraded by glutathione peroxidase (GPx) or catalase (Cat). Superoxide can also react with nitric oxide (NO) or arachidonic acid (AA) to form peroxynitrite ($ONOO^-$) or isoprostanes, respectively. In addition to other signaling effects, H_2O_2 can activate Nox, resulting in further production of superoxide. Lastly, H_2O_2 can be converted to hydroxyl radical (a highly reactive molecular species) in the presence of iron (Fe^{2+})

diffusion limited reaction) that the local concentration of superoxide is a key determinant of the biological half-life (or bioactivity) of NO [1]. Because NO is a potent vasodilator, the reaction of NO with superoxide has two major consequences. First, it results in loss of the chronic vasodilator influence that NO exerts on vascular tone normally, with resulting vasoconstriction. Second, peroxynitrite is formed as a result of this reaction (Fig. 6.1). In addition to peroxynitrite, isoprostanes are formed from the peroxidation of arachidonic acid by oxygen-derived free radicals including superoxide (Fig. 6.1). As will be discussed below, both peroxynitrite and isoprostanes have their own effects on vascular tone.

What Are the Major Sources of ROS in the Cerebral Circulation?

There are multiple sources of ROS within vascular cells including mitochondria, cyclooxygenase (COX), and NADPH oxidases, among others [1, 11, 12]. Cerebral blood vessels, particularly brain endothelial cells, contain a relatively high density of mitochondria [13]. Mitochondria are a major source of superoxide that is generated during production of ATP (oxidative phosphorylation) [14].

It has been known for many years that COX is an important source of superoxide in the cerebral circulation [15]. Arachidonic acid is the precursor for many prostanoids, producing vasodilation or vasoconstriction depending on which pathway of arachidonic acid metabolism predominates, which prostanoids are produced, and what prostanoid receptors are present locally. In relation to vascular biology, the COX pathway has been widely studied. Superoxide is produced during COX activity (i.e., the conversion of arachidonic acid to prostaglandin H_2). In this context, arachidonic acid elicits COX-dependent formation of superoxide in cerebral arteries and the brain microcirculation [15–17]. In the presence of reduced activity of superoxide dismutase (SOD) activity, steady-state increases in superoxide levels in cerebral arteries in response to arachidonic acid are enormously augmented [15], providing an example of how endogenous SOD activity plays a major role in limiting increases in local levels of superoxide.

In recent years, a substantial effort has been made to define the importance of NADPH oxidases in vascular biology, including in the cerebral circulation (Table 6.1) [11, 18, 19]. The NADPH oxidase family of enzymes has emerged as a major source of ROS in the vasculature. There are multiple isoforms of NADPH oxidase, defined based on which enzymatic subunit is responsible for ROS formation ("Nox" subunit) [11, 18, 19]. The catalytic domain of NADPH oxidase resides in the Nox subunit. A number of other membrane-bound and cytosolic subunits are required for ROS generation by Nox enzymes; however, the number and identity of these subunits vary depending on the isoform. Within vascular cells, the isoforms of interest are Nox1, Nox2, Nox4, and Nox5 (Table 6.1) [11]. Due mainly to which genetic models have been available, most insight to date has been obtained regarding the functional importance of the Nox2-containing isoform of NADPH oxidase.

Table 6.1 Expression of Nox isoforms in cerebral blood vessels

Nox isoform	Cell type	Method
Nox1	Endothelium Vascular muscle adventitia	Immunohistochemistry [143]
	N/A	RT-PCR [143, 144]
	N/A	Western blotting [143, 145–147]
Nox2	Endothelium	Immunofluorescence [4, 148]
	Endothelium	Electron microscopy [58]
	Adventitia	Immunofluorescence [4, 148]
	N/A	RT-PCR [135, 143, 144]
	N/A	Western blotting [4, 56, 147–149]
Nox4	Endothelium Vascular muscle adventitia	Immunofluorescence [150]
	N/A	RT-PCR [143, 144]
	N/A	Western blotting [3, 145, 147]

N/A not applicable, *RT-PCR* reverse transcription polymerase chain reaction

In relation to specific ROS, Nox1, Nox2, and Nox5 produce superoxide, while Nox4 is thought to produce hydrogen peroxide [11]. Expression of Nox4 is relatively abundant, normally being present in all vascular cells [11]. In relation to the current discussion, it is noteworthy that Nox4 is expressed at relatively high levels in cerebral arteries [3, 11]. Unfortunately, little is still known regarding the functional importance of this isoform. Nothing is known regarding the presence or impact of Nox5 in brain vessels, in part because this isoform is not expressed in the rodent genome [11, 18, 19], which includes the most commonly used experimental models. Nox5 is expressed in human blood vessels [20], but its presence and impact in cerebrovasculature remains to be determined.

Under select conditions, another potential source of superoxide in the vasculature is endothelial NO synthase (eNOS) [12]. For example, after oxidation of the enzyme cofactor tetrahydrobiopterin, eNOS can reduce molecular oxygen, instead of its normal substrate L-arginine, producing superoxide rather than NO. While the possibility of eNOS "uncoupling" is often discussed, few studies of the cerebral circulation have provided definitive evidence that eNOS is a functionally important source of superoxide. One exception is the work of Katusic et al, who have shown that reduced bioavailability of tetrahydrobiopterin causes eNOS uncoupling and eNOS-dependent superoxide production. These changes were sufficient to impair eNOS-cGMP-dependent signaling in the cerebral microvasculature [21, 22].

Direct Effects of ROS and Related Molecules on Vascular Tone

Several approaches have been used to test the effects of ROS on vascular tone. A common strategy when quantifying cellular effects of hydrogen peroxide has been direct application of the molecule to blood vessels or vascular cells. For other relevant ROS, the biochemical properties of the molecules (e.g., half-life, reactivity, etc.) make such a simple approach extremely challenging. Instead, local generation of ROS using enzymatic preparations such as xanthine/xanthine oxidase has been a relatively common method [23]. A more selective approach includes exposing the cerebral vasculature to NADPH or NADH (substrate for NAPDH oxidase) [24, 25]. The precise mechanism by which application of NADH/NADPH increases vascular ROS production is unclear as NADH/NADPH are large, charged molecules that are unlikely to diffuse across the cell membrane to bind the substrate binding domain of the Nox subunit [26]. To our knowledge, no specific NADH/NADPH transmembrane transport protein has been identified. Despite the current uncertainty regarding the mechanism by which application of NADH/NADPH increases vascular ROS, numerous studies have demonstrated that this method can increase NADPH oxidase activity.

In relation to superoxide, previous studies have described diverse effects, including both vasoconstriction and vasodilation depending on the species and experimental conditions. For example, superoxide elicits dilation of cerebral arterioles in response to NADH/NADPH [24, 25]. Superoxide has also been shown to cause dilation of cerebral microvessels when produced by the ROS-generating enzyme xanthine oxidase [23]. In contrast, superoxide-mediated contraction of cerebral arteries has been described when treated with A23187 or NADH [24, 27]. Vasoconstriction to ROS may involve activation of protein kinase C and L-type calcium channels [28]. Responses to superoxide appear to be concentration dependent, with vasodilation at low concentrations and vasoconstriction at high concentrations [29]. Hydrogen peroxide is chemically more stable and more cell membrane-permeable than superoxide, and effects of hydrogen peroxide on vascular tone tend to be more consistent in the literature. Hydrogen peroxide most commonly produces dilation of cerebral arteries and almost always dilates the brain microcirculation [3, 23, 29, 30]. When vasodilator effects of either superoxide or hydrogen peroxide are present, they are typically mediated by activation of various potassium channels in vascular muscle, particularly calcium-activated potassium channels [23, 30–32]. Lastly, it is important to note that hydrogen peroxide can stimulate NADPH oxidase in vascular cells, further amplifying ongoing oxidant-dependent mechanisms (Fig. 6.1) [33].

Like ROS, peroxynitrite can produce either vasodilation or vasoconstriction [29, 32]. For example, low concentrations of peroxynitrite dilate brain microvessels via potassium channel-dependent mechanisms [23, 29, 32]. In contrast, peroxynitrite can constrict cerebral arteries via inhibition of basal activity of calcium-dependent potassium channels [34, 35]. In a similar manner to superoxide and hydrogen peroxide, these opposing influences on vascular tone may reflect concentration-dependent effects. Peroxynitrite can also affect vascular tone through indirect effects including impairment of endothelial function and neurovascular coupling [29, 36]. For these latter effects, Nox2-containing NADPH oxidase is an important source of superoxide contributing to peroxynitrite formation [36]. Peroxynitrite can also inhibit myogenic tone in cerebral arteries via a mechanism that appears to involve nitrosylation of F-actin in vascular muscle [37]. Impairment of regulation of vascular tone by peroxynitrite can also involve activation of poly (ADP-ribose) polymerases (PARP). Excessive activation of PARP contributes to endothelial dysfunction in models of disease and aging [1, 36, 38, 39]. Additional effects of peroxynitrite that can affect vascular tone include reducing activity of prostacyclin synthase and SOD as a result of nitration of tyrosine residues [1]. Recent work suggests detrimental effects of PARP on endothelial cells can be mediated by downstream activation of TRPM2 channels [40].

Isoprostanes have relatively potent effects on vascular tone. The most studied of these molecules are the series of F_2-isoprostanes which produce constriction of cerebral arteries and microvessels via a thromboxane-like receptor (Fig. 6.1) [29, 41]. In relation to cerebrovascular disease, isoprostanes have emerged as a highly reliable marker of oxidative stress [29, 41] in addition to their direct effects on vascular tone. We discuss their potential contribution to endothelium-dependent contraction below.

Considerable effort has been directed at defining the importance of endogenously generated ROS in models of vascular disease and stroke. In addition to quantifying changes in local levels of these molecules in the vasculature, a common approach involves testing the functional impact of ROS using pharmacological, molecular, or genetic interventions to inhibit formation of ROS. Equally common are studies designed to quantify the influence of ROS by reducing their levels (scavenging the molecules) in order to inhibit their effects. Although pharmacological scavengers like exogenous SOD, tempol (1-oxyl-2,2,6,6-tetramethyl-4-hydroxypiperidine), or MnTBAP [manganic (I–II)meso-tetrakis (4-benzoic acid) porphyrin] are used most commonly, genetic models with altered expression of antioxidant enzymes have also been utilized to address these questions.

Endothelial Function

One of the most studied areas of vascular biology involves how endothelial cells regulate tone of underlying vascular muscle (Fig. 6.2). This collective process continues to receive broad attention from both basic researchers and clinicians because of its key role in the control of organ vascular resistance and its diverse impact, not only on vascular muscle, but also other cell types including circulating immune cells, platelets, neurons, and glia [2, 42–45].

Endothelium is a key site of end-organ damage. The phrase endothelial dysfunction is used commonly to describe endothelial-based abnormalities that promote vasoconstriction, inflammation, increased permeability, atherosclerosis, and thrombosis. Endothelial dysfunction is generally considered to be the earliest event in the initiation of vascular disease, but also plays a key role throughout the disease process. Both genetic and clinical data have established that the vasomotor component of endothelial dysfunction is predictive of cardiovascular events, including stroke [43–46]. In brain, the consequences, of endothelial dysfunction include atherosclerosis, large and small vessel disease, stroke, dementia, and blood–brain

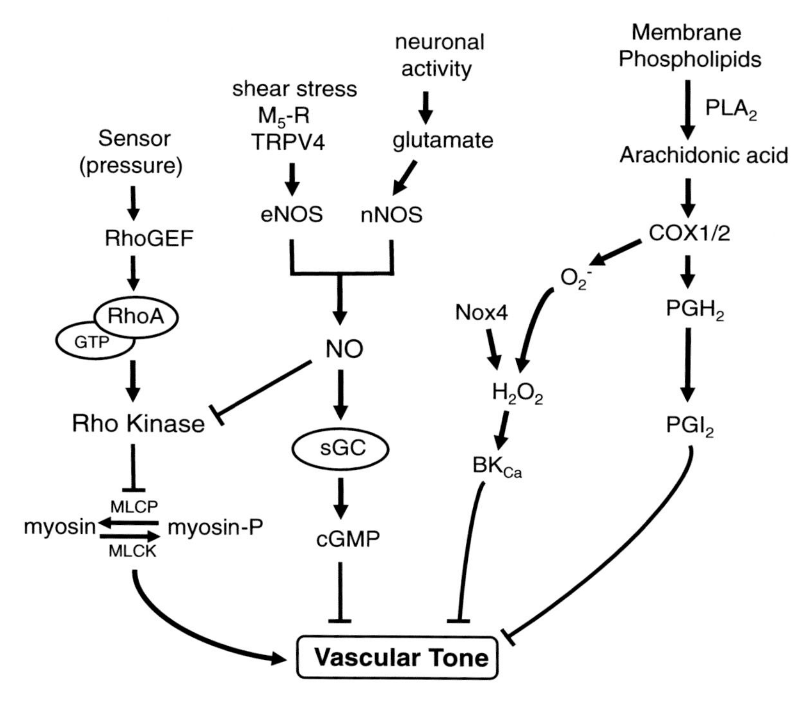

Fig. 6.2 Select signaling pathways that affect vascular tone under normal conditions. Increases in intravascular pressure (transmural pressure) increase activity of RhoA and Rho kinase by first stimulating guanine nucleotide exchange factor (GEF). The contractile state of vascular muscle is determined by the phosphorylation state of myosin light chain (myosin versus myosin-P). Both endothelial (eNOS) and neuronal (nNOS) NO synthase produce NO, which signals through its major target, soluble guanylate cyclase (sGC) and cGMP. Activators of eNOS include acetylcholine via the type 5 muscarinic receptor (M_5-R) [140], activation of the V4 subtype of transient receptor potential channels (TRPV4) [80, 141] or shear stress [142]. Release of the neurotransmitter glutamate activates nNOS in target neurons. Cyclooxygenase (COX) metabolizes arachidonic acid to prostaglandin H_2, which can be converted to several prostanoids including prostacyclin (PGI_2). During this process, COX also produces superoxide anion (O_2^-) which is converted to hydrogen peroxide (H_2O_2) by superoxide dismutases (SOD). Hydrogen peroxide, which can also be formed by Nox4, activates potassium channels (BK_{Ca} channels) in vascular muscle, producing membrane hyperpolarization and relaxation. *MLCP* myosin light chain phosphatase, *MLCK* myosin light chain kinase, *PLA_2* phospholipase A_2

barrier abnormalities. Functional changes in endothelial cells contribute to reductions in resting blood flow (hypoperfusion), impairment of vasodilator responses, and subsequent cellular injury.

Endothelium communicates with target cells via mechanisms that include the production and extracellular release of signaling molecules (Fig. 6.2) as well the transfer of electrical signals from cell to cell (e.g., endothelium-dependent hyperpolarization) [2, 31, 45]. The latter of these mechanisms utilizes gap junctions between adjacent endothelial cells or between endothelium and vascular muscle (myoendothelial junctions) [47].

Of the endothelium-derived molecules that are vasoactive, the impact of NO produced by eNOS has received the most study. Through this pathway, endothelial cells normally exert a tonic vasodilator influence on large and small vessels on the brain surface as well as arterioles within the parenchyma (Fig. 6.2). Several lines of evidence support this concept including acute effects of inhibitors of NO synthase on vascular tone and cerebral blood flow [2, 31, 42]. Unlike upstream arteries and arterioles which include layers of vascular muscle that generate myogenic tone, capillaries do not have intrinsic resting tone. However, capillaries can undergo changes in diameter as a result of alterations in intravascular pressure (distending pressure) or as a result of the influence of local pericytes. Pericytes, contractile cells that extend processes along and around capillaries, are also responsive to vasoactive stimuli and thus can influence capillary diameter [48, 49]. In cerebellar slice preparations for example, NO can increase the diameter of capillaries via effects on pericytes [50]. How important pericyte-dependent changes in capillary diameter are in relation to changing vascular resistance in vivo is poorly understood at present.

The primary molecular target of NO in vascular muscle is soluble guanylate cyclase [51–53]. When activated, this enzyme synthesizes the second messenger cGMP from GTP (Fig. 6.2) [2, 31, 42]. Downstream effects of cGMP on vascular tone are well described and are mediated by several mechanisms including activation of cGMP-dependent protein kinase I with resulting reductions in intracellular calcium [52]. Through this eNOS-NO-cGMP-dependent mechanism, endothelial cells influence resting vascular tone and mediate vasodilator responses to a diverse group of physiological, pharmacological, and mechanical stimuli. Thus, endothelium-derived NO is the mediator of cerebrovascular responses to neurotransmitters, metabolic factors, shear stress, products released by other cells (platelets and astrocytes), and therapeutic agents (Fig. 6.2) [2, 54].

Table 6.2 provides a summary of studies that have implicated oxidative stress as a mediator of endothelial dysfunction in models of vascular disease and in cerebral arteries from people. The models in which such effects have been described are diverse and include common risk factors for vascular disease (eg, hypertension, diabetes and obesity, hypercholesterolemia, and aging), models of stroke and small vessel disease, as well as genetic models with altered expression of select genes. The majority of these studies have concluded that superoxide was a key mediator of mechanisms responsible for oxidative stress. For a subset of that group, Nox2-containing NADPH oxidase was found to play an essential role. Key Nox-2 dependent effects have been described in models of obesity, hypertension, Alzheimer's disease, stroke, and aging [18, 23, 36, 55–62]. In other cases, hydrogen peroxide, hydroxyl radical, or peroxynitrite has been implicated. For most of these studies, it is the NO-component of endothelium-dependent regulation of vascular tone that has either been demonstrated, or assumed to be affected by oxidative stress [38, 63–65].

The majority of insight in this area has been obtained by studying cerebral arteries (either in vivo or in vitro), the cerebral (pial) microcirculation (almost always in vivo), or measurements of local cerebral blood flow. In contrast to pial arterioles,

Table 6.2 Studies examining the impact of oxidative stress on endothelium-dependent responses in the cerebral circulation

Model	Vasculature
ACE2 deficiency	Basilar artery [135]
Activated neutrophils	Middle cerebral artery [151]
Acute hypertension	Pial arterioles [152]
Aging	Basilar artery [38] Middle cerebral artery [63] Cerebral blood flow [60] Pial arterioles [153]
Alcohol	Pial arterioles [154, 155] Basilar artery [156]
Aldosterone	Basilar artery [55]
Alzheimer's disease	Cerebral blood flow [40, 79] Cerebral blood flow [62] Middle cerebral artery [64, 157] Posterior cerebral artery [80]
Ang II-dependent hypertension	
Non-pressor Slow-pressor Pressor Genetic (R^+A^+)	Basilar artery [118] Cerebral blood flow [158, 159] Basilar artery [18, 160] Basilar artery [161]
Aortic banding	Pial arterioles [6]
Cerebral cavernous malformation	Middle cerebral artery [162]
Dehydration	Cerebral blood flow [163]
Diabetes/metabolic syndrome	
Streptozotocin (type 1)	Pial arterioles [146] Basilar artery [164]
db/db (type 2) Tally Ho (type 2) OLETF (type 2) Zucker obese Zucker obese	Pial arterioles [78] Pial arterioles [165] Basilar artery [166] Basilar artery [167] Middle cerebral artery [113]
Diet-induced obesity	Pial arterioles [59]
DOCA-salt	Middle cerebral artery [168] Pial arterioles [168]
Fluid percussion injury	Pial arterioles [169]
Hypercholesterolemia	Middle cerebral artery [149] Pial arterioles [170]
Hyperhomocysteinemia	Pial arterioles [171]
Hypoxia/reoxygenation	Posterior cerebral artery [172]
Intermittent hypoxia	Cerebral blood flow [173]
Ischemia/reperfusion	Pial arterioles [84] Middle cerebral artery [56] Cerebral blood flow [174]
Lipopolysaccharide	Middle cerebral artery [175]

(continued)

Table 6.2 (continued)

Model	Vasculature
Nicotine	Basilar artery [176] Pial arterioles [145]
PPARγ interference	
Systemic (all cells)	Pial arterioles [129] Basilar artery [129]
Endothelial specific	Basilar artery [130]
Simulated microgravity	Basilar artery [177]
SOD deficiency	
SOD1	Basilar artery [85] Pial arterioles [7, 114]
SOD2	Basilar artery [115] Pial arterioles [115]
SOD3	Pial arterioles [116]

much less is known regarding endothelial function in parenchymal arterioles, a key segment of the vasculature in relation to small vessel disease [66]. Endothelial dysfunction occurs in parenchymal arterioles in models of disease [67–69], but the role of oxidative stress and ROS in these changes has received little attention. One study provides evidence that superoxide accounts for impaired acetylcholine-induced dilation of arterioles in brain slices during hyperglycemia [69].

In addition to its immediate effects, loss of endothelium-derived NO signaling can affect vascular tone through other mechanisms. First, measurements of arachidonic acid metabolism have shown that inhibition of eNOS (NO production) increases vascular production of vasoconstrictor isoprostanes, including 8-iso-PGF_2 [70]. Activation of thromboxane receptors by endothelial-derived isoprostanes may account for endothelium-dependent contraction described in cerebral arterioles in models of hypertension and diabetes (Fig. 6.3) [71–73]. Second, low grade inflammation is a common feature of many forms of vascular disease. eNOS-derived NO normally decreases activation of nuclear factor κB (NF-κB) and expression of pro-inflammatory genes [74]. Loss of this protective mechanism promotes vascular inflammation, which can affect both vasoconstrictor and vasodilator responses. Third, activity of Rho kinase is a key determinant of vascular tone [75–77]. Because NO and cGMP normally inhibit activation of Rho kinase in vascular muscle (Fig. 6.2), reductions in NO increase activity of the enzyme (Fig. 6.3) [75–77]. It is noteworthy that inhibition of NO production rapidly increases the influence of Rho kinase on microvascular tone in brain [78].

To what extent do ROS influence endothelium-dependent regulation of vascular tone in the absence of vascular disease? In relation to NO-mediated responses, the answer appears to be relatively little. For example, dilation of cerebral blood vessels to acetylcholine, ATP, and other endothelium-dependent agonists are not affected by SOD and catalase in several species, suggesting no influence of superoxide or hydrogen peroxide on vascular tone [79–84]. Similarly, responses of cerebral arteries to acetylcholine are not affected in transgenic mice overexpressing the cytosolic

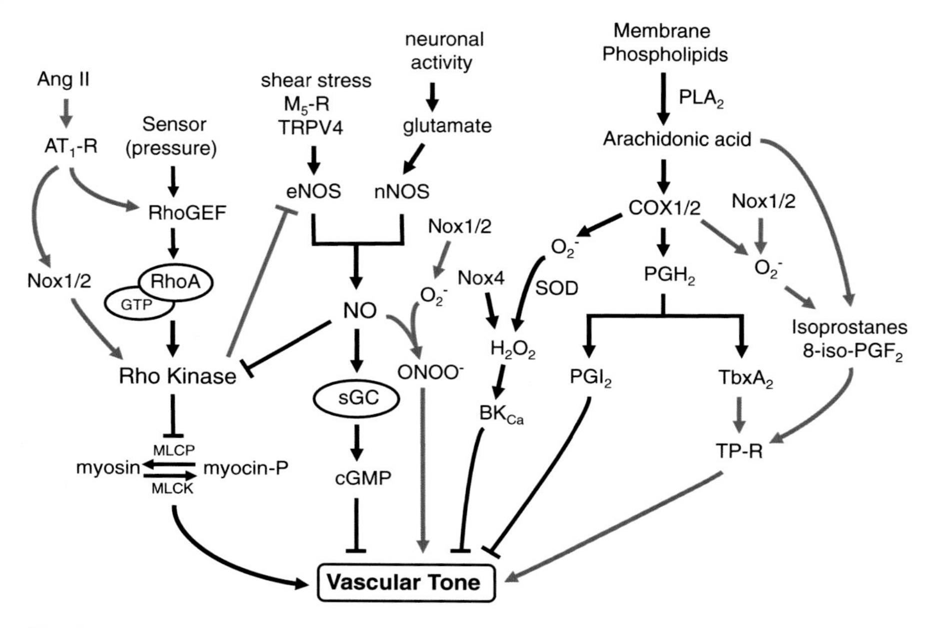

Fig. 6.3 Select pathways that influence the regulation of vascular tone in disease. This schematic builds on the concepts presented in Fig. 6.2 noting disease-related elements in *red*. Angiotensin II (Ang II), acting via the AT_1-receptor (AT_1-R) increase activity of Rho kinase via effects on RhoGEF or by promoting Nox-dependent formation of ROS. In addition to increasing vascular tone directly, Rho kinase inhibits production of NO by eNOS. Superoxide (O_2^-), produced by Nox or other enzymatic sources, readily reacts with NO to form peroxynitrite ($ONOO^-$), which can contract vascular muscle. Superoxide can also react with arachidonic acid results in generation of isoprostanes. Both isoprostanes and thromboxane A_2 ($TbxA_2$), produced via the COX pathway, activate prostanoid TP receptors (TP-R) to produce contraction of vascular muscle

form of glutathione peroxidase, which degrades hydrogen peroxide [85]. Dilation of normal cerebral arteries to an activator of transient receptor potential vanilloid 4 (TRPV4) channels, which underlies a substantial portion of the response to acetylcholine, is also not affected by SOD or catalase [80].

Endothelium-dependent, but NO-independent mechanisms, also affect vascular tone. An example of this concept is the vasodilation mediated by endothelium-dependent hyperpolarization or release of endothelium-derived hyperpolarizing factors (EDHF) [47, 86]. Both pharmacological approaches and studies of transgenic mice overexpressing glutathione peroxidase support the concept that hydrogen peroxide functions as an EDHF (Fig. 6.2) [16, 30, 85]. This signaling mechanism accounts for microvascular effects of bradykinin and effects of arachidonic acid on both cerebral arteries and arterioles (Fig. 6.2) [16, 30, 85]. The immediate source of hydrogen peroxide underlying this effect appears to be the cytosolic form of SOD (SOD1), which converts superoxide to hydrogen peroxide [85].

Neurovascular Coupling

Among the mechanisms that regulate cerebral perfusion, neurovascular coupling (or functional hyperemia) is an important adaptive response [87, 88]. With increases in cellular activity, the brain requires proportionally more blood flow [87, 88]. Normally this change in local demand is closely matched by rapid increases in local blood flow [87, 88]. The term neurovascular coupling reflects the concept that in response to neural activation, neurons, glia, and endothelium produce signals that communicate with the vasculature resulting in vasodilation, including conducted or flow-dependent dilation of upstream vessels that supply the region where the activation occurs [87–89]. Neurovascular coupling ensures adequate delivery of oxygen, glucose, and other nutrients during cellular activation, as well as efficient removal of metabolic by-products. The conducted dilation of upstream vessels also contributes to maintenance of local microvascular pressure (local perfusion pressure) [90, 91].

Mechanisms that mediate and modulate neurovascular coupling have received increasing attention in recent years. Chronic impairment of neurovascular coupling is thought to contribute to the decline in brain function that occurs with aging or dementia (including small vessel disease and Alzheimer's disease) [88, 92]. Similarly, loss of effective neurovascular coupling is thought to play a role in the decline in cognitive function that occurs in the presence of cardiovascular risk factors like hypertension and diabetes [25, 58, 60, 64, 65, 93, 94]. Stroke impairs neurovascular coupling in both humans and experimental models [95]. In relation to mechanisms, ROS have been implicated in impaired neurovascular coupling in models of Alzheimer's disease, hypertension, and aging [25, 36, 40, 58, 60, 65, 96].

The contribution of specific molecules and signaling pathways in neurovascular coupling varies regionally. In regions like the somatosensory cortex, there is controversy regarding the mediators of functional hyperemia [97]. In relation to the current topic, one molecule that has been implicated in functional hyperemia in the cerebral cortex is adenosine [98]. Some work has suggested that adenosine-induced cerebral vasodilation is mediated by ROS [99]. For example, adenosine produced dilation of isolated cerebral arteries, an effect that was mediated by A_{2A} and A_{2B} adenosine receptors and inhibited by SOD and catalase [99]. Such findings would imply that a component of functional hyperemia (an adenosine-dependent component) may also require formation of ROS. While an interesting concept, supporting evidence for such a mechanism in vivo seems lacking. Both pharmacological and genetic approaches, including the use of Nox2 deficient animals indicate that cerebrovascular responses to adenosine or whisker stimulation (functional hyperemia) are normally independent of superoxide [25, 58, 60–62, 96].

Chemoregulation

The partial pressure of CO_2 in arterial blood ($PaCO_2$) is a powerful regulator of cerebral vascular resistance and thus blood flow in brain [100]. Hypercapnia (increases in arterial $PaCO_2$) causes marked vasodilation. Conversely, reductions in

$PaCO_2$ (hypocapnia) sharply increases cerebrovascular resistance, thus reducing cerebral blood flow [100]. Work by Kontos and others has shown that the changes in vascular tone that occur with alterations in $PaCO_2$ are due to changes in extracellular pH, not changes in $PaCO_2$ per se [100, 101].

While effects of hypercapnia on cerebral blood flow have been widely studied, the impact of ROS on these responses has not been. Some studies have shown that increases in diameter of cerebral arterioles and cerebral blood flow during hypercapnia are not affected by scavengers of ROS or a cell-permeable inhibitor of NADPH oxidase (gp91ds-tat) [60, 79, 102–104]. It is not uncommon for models of vascular disease to exhibit superoxide-mediated endothelial dysfunction and impaired neurovascular coupling but normal responses to hypercapnia [58, 79, 94, 105]. Such results could imply that vascular effects of hypercapnia are not as susceptible to effects of oxidative stress. In circumstances where cerebrovascular responses to hypercapnia are impaired, there is evidence both for and against a contributing role for ROS, depending on the model. For example, in transgenic mice expressing Swedish, Iowa, and Dutch mutations of amyloid precursor protein (a model of cerebral amyloid angiopathy), an SOD mimetic (MnTBAP) improved vascular responses to acetylcholine and whisker stimulation, but not responses to hypercapnia [96]. In contrast, SOD restored cerebral blood flow changes during hypercapnia to normal in a model of hyperhomocysteinemia [104].

Myogenic Responses and Autoregulation

The term autoregulation refers to the ability of the brain to maintain cerebral blood flow relatively constant over a range of perfusion pressures [106]. Multiple mechanisms, contributing in an integrated fashion, underlie this well described feature of the cerebral circulation. In vivo, myogenic responses in cerebral arteries and cerebral arterioles (both on the pial surface and within the parenchyma) are a key component of autoregulation (vasodilation to low perfusion pressure and vasoconstriction to elevated perfusion pressure) [106, 107]. Although cerebrovascular disease is often accompanied by alterations in myogenic responses in isolated vessels and in autoregulation in vivo, mechanisms that control these changes are still relatively poorly understood.

Kontos and others demonstrated that very large increases in arterial pressure (acute hypertension) produce dilation of cerebral arteries and arterioles that was at least partly mediated by ROS [108, 109]. In contrast, the possibility that ROS participate in regulation of CBF within the autoregulatory range has not been studied extensively. Recently, it was suggested that ROS contribute to enhanced myogenic tone in isolated cerebral arteries in response to increased intravascular pressure [110]. That study described elevated superoxide and myogenic tone in response to increased intravascular pressure, with the increases in vascular tone being inhibited by scavengers of superoxide. NADPH oxidase may be a source of superoxide mediating this effect as treatment with an inhibitor of the oxidase reduces myogenic tone

[111]. In contrast, other studies reported no effect of ROS scavengers on myogenic tone in normal cerebral arteries using similar approaches [112, 113]. Unlike these findings in normal arteries, scavenging superoxide reduced the augmented myogenic tone that is present in a model of obesity [112, 113]. In vivo, there are many studies where scavengers of superoxide and/or hydrogen peroxide do not affect resting diameter of cerebral arterioles under normal conditions, even though these vessels have substantial myogenic tone. Thus, the overall importance of ROS in relation to myogenic tone and myogenic responses is not entirely clear. ROS may have little influence under normal conditions, but may affect myogenic tone in models of disease or injury, including hypertension.

Loss of Mechanisms That Protect Against Oxidative Stress

Recent studies have highlighted a number of endogenous mechanisms that protect vascular cells, including in the cerebral circulation. Loss of these protective molecules or mechanisms promotes oxidative stress and influence vascular tone. In relation to steady-state conditions, increases in local levels of vascular ROS can occur as a result of increased production or reduced degradation by relevant anti-oxidant enzymes [1].

A family of antioxidant enzymes is normally expressed within the vessel wall. There are three isoforms of SOD in different subcellular compartments, each converting superoxide to hydrogen peroxide (Figs. 6.1 and 6.3). The isoform of SOD that is the greatest contributor to total SOD activity in the vasculature is cytosolic SOD1 (CuZn-SOD) [1, 114]. Pharmacological inhibition of Cu-containing SODs [15, 85], or selective genetic deficiency in SOD1 [114], increase vascular superoxide levels, impair endothelial function in cerebral arteries and arterioles, and augment vasoconstrictor responses [7, 85, 114]. SOD2 (manganese SOD) is localized in mitochondria, one of the most important cellular sources of superoxide. Genetic deficiency in SOD2 enhances constrictor responses of cerebral arteries and produces superoxide-mediated endothelial dysfunction in brain microvessels [115]. Unlike SOD1 and SOD2, genetic deficiency in SOD3 (extracellular SOD) has no effect on resting tone or endothelial function [116]. In addition to these baseline effects, genetic deficiency in any of the SODs predisposes to augmented endothelial dysfunction when challenged with risk factors for vascular disease, including aging and angiotensin II [39, 116–119]. Similar to the findings with SOD1 and SOD2, partial genetic deficiency in glutathione peroxidase 1 (Fig. 6.1), greatly augments effects of angiotensin II on endothelial function [120].

While not traditionally considered antioxidants, there are other molecules that exert important anti-oxidant effects in the cerebrovasculature. Peroxisome proliferator-activated receptor-γ (PPARγ) is a ligand-activated transcription factor expressed in many cell types including those within the vasculature [121–123]. Recent studies using genetic approaches or synthetic activators of PPARγ (e.g., thiazolidinediones, TZDs) have revealed protective effects of PPARγ in models of disease and in patients with atherosclerosis [121–125]. TZDs have beneficial effects on both endothelial function and regulation of vascular tone in models of Alzheimer's disease and hypertension [64, 121–123, 125, 126]. For example, vascular dysfunction in the brain in aged mice overexpressing amyloid-precursor protein, a model of Alzheimer's disease, was restored toward normal following TZD treatment [64]. These findings suggest that pharmacological activation of PPARγ can have protective effects on vascular function.

Dominant negative mutations in the ligand-binding domain of PPARγ inhibit transcriptional activity of wild-type PPARγ [127, 128]. Effects exerted by these mutations are opposite to those produced by TZDs [127, 128]. The recent development of mice expressing dominant negative forms of PPARγ has provided the opportunity to begin to define the importance of wild-type PPARγ driven by endogenous ligands (without the need for TZD treatment).

Heterozygous knockin mice expressing one of these PPARγ mutations (P465L) in all cells exhibit oxidative stress and endothelial dysfunction in both cerebral arteries and arterioles [129]. In transgenic mice expressing dominant negative PPARγ under control of an endothelial-specific promoter, vascular function appears normal under baseline conditions, but cerebral arteries are predisposed to superoxide-mediated endothelial dysfunction when mice are challenged with a high-fat diet [130]. In contrast, genetic interference with PPARγ specifically in vascular muscle impaired NO-mediated responses in both large and small cerebral vessels, while augmenting select vasoconstrictor responses [51]. Interestingly, and in contrast to the endothelial specific model, these changes did not appear to involve oxidant-dependent mechanisms. Overall, this work indicates that interference with PPARγ has pronounced effects in the cerebral circulation, but the contribution of oxidative stress to these effects varies depending on the cell type involved.

The renin-angiotensin system generally promotes vascular disease, primarily via effects of angiotensin II acting through the AT1 receptor (Fig. 6.3) [2, 131–133]. In contrast to angiotensin II, this system can exert beneficial effects and antagonize effects of Ang II via actions of angiotensin 1–7 produced by angiotensin-converting enzyme 2 (ACE2) [134]. It is thought that ACE2 and angiotensin 1–7 acting via mas receptors provide a mechanistic balance which counters the pro-oxidant effects of angiotensin II acting via AT1 receptors [134]. Consistent with this concept, genetic deletion of ACE2 produced endothelial dysfunction in cerebral arteries via a superoxide-dependent mechanism [135]. In addition, ACE2 deficiency augmented endothelial dysfunction during aging [135].

Conclusions

In this chapter, we have outlined a variety of effects of ROS and oxidative stress in relation to regulation of cerebral vascular tone. With a few exceptions, ROS appear to exert little influence on vascular tone under normal or baseline conditions. In this regard, the vast majority of studies found no evidence for ROS affecting resting vascular tone (or vascular resistance), endothelium-dependent vasodilation, or neurovascular coupling. Less data is available regarding the impact of ROS on myogenic tone or autoregulation. In contrast to baseline conditions, these molecules have substantial effects on vascular tone in models of disease and with aging. We also summarized enzymatic sources of ROS as well as select endogenous protective molecules that have a substantial influence on vascular tone through oxidant-dependent mechanisms.

In support of the concepts outlined above on the impact of ROS in the cerebral circulation, are related findings from peripheral blood vessels in humans [136–138]. Despite what might be perceived as overwhelming experimental evidence implicating ROS in models of vascular disease, large scale clinical trials testing the effectiveness of antioxidants (vitamin supplementation) in reducing cardiovascular events, including stroke, have largely been disappointing (e.g., the Heart Protection Study) [139]. Such an outcome may potentially be due to the inherently poor antioxidant capacity of vitamins and their inability to access all relevant subcellular compartments. Therefore, pharmacological approaches using more efficacious and selective antioxidants may be needed to impact cardiovascular events. In addition, therapeutic approaches that target major enzymatic sources of ROS in the vasculature (e.g., NADPH oxidase) may result in better outcomes.

Acknowledgement Work summarized in this chapter was supported by research grants from the National Institute of Health (NS-096465, NS-24621, HL-62984, and HL-113863), the Department of Veteran's Affair's (BX001399), and the Fondation Leducq (Transatlantic Network of Excellence on the Pathogenesis of Cerebral Small Vessel Disease). TMD was the recipient of an Overseas Post-doctoral Fellowship from the National Health and Medical Research Council of Australia (1053786).

References

1. Faraci FM, Didion SP. Vascular protection: superoxide dismutase isoforms in the vessel wall. Arterioscler Thromb Vasc Biol. 2004;24:1367–73.
2. Faraci FM. Protecting against vascular disease in brain. Am J Physiol. 2011;300:H1566–82.
3. Miller AA, Drummond GR, Schmidt HH, Sobey CG. NADPH oxidase activity and function are profoundly greater in cerebral versus systemic arteries. Circ Res. 2005;97:1055–62.
4. Miller AA, Drummond GR, De Silva TM, Mast AE, Hickey H, Williams JP, et al. NADPH oxidase activity is higher in cerebral versus systemic arteries of four animal species: role of Nox2. Am J Physiol. 2009;296:H220–5.
5. Chan SL, Baumbach GL. Deficiency of Nox2 prevents angiotensin II-induced inward remodeling in cerebral arterioles. Front Physiol. 2013;4:133.
6. Chan SL, Baumbach GL. Nox2 deficiency prevents hypertension-induced vascular dysfunction and hypertrophy in cerebral arterioles. Int J Hypertens. 2013;2013:793630.
7. Baumbach GL, Didion SP, Faraci FM. Hypertrophy of cerebral arterioles in mice deficient in expression of the gene for CuZn superoxide dismutase. Stroke. 2006;37:1850–5.
8. Starke RM, Chalouhi N, Ali MS, Jabbour PM, Tjoumakaris SI, Gonzalez LF, et al. The role of oxidative stress in cerebral aneurysm formation and rupture. Curr Neurovasc Res. 2013;10:247–55.

9. Wood KC, Hebbel RP, Granger DN. Endothelial cell NADPH oxidase mediates the cerebral microvascular dysfunction in sickle cell transgenic mice. FASEB J. 2005;19:989–91.
10. Freeman LR, Keller JN. Oxidative stress and cerebral endothelial cells: regulation of the blood-brain-barrier and antioxidant based interventions. Biochim Biophys Acta. 1822;2012: 822–9.
11. Drummond GR, Sobey CG. Endothelial NADPH oxidases: which NOX to target in vascular disease? Trends Endo Metab. 2014;25:452–63.
12. Leopold JA, Loscalzo J. Oxidative risk for atherothrombotic cardiovascular disease. Free Radic Biol Med. 2009;47:1673–706.
13. Oldendorf WH, Cornford ME, Brown WJ. The large apparent work capability of the blood-brain barrier: a study of the mitochondrial content of capillary endothelial cells in brain and other tissues of the rat. Ann Neurol. 1977;1:409–17.
14. Busija DW, Katakam PV. Mitochondrial mechanisms in cerebral vascular control: shared signaling pathways with preconditioning. J Vasc Res. 2014;51:175–89.
15. Didion S, Hathaway C, Faraci F. Superoxide levels and function of cerebral blood vessels after inhibition of CuZn-SOD. Am J Physiol. 2001;281:H1697–703.
16. Kontos HA, Wei EP, Kukreja RC, Ellis EF, Hess ML. Differences in endothelium-dependent cerebral dilation by bradykinin and acetylcholine. Am J Physiol. 1990;258:H1261–6.
17. Kontos HA, Wei EP, Povlishock JT, Christman CW. Oxygen radicals mediate the cerebral arteriolar dilation from arachidonate and bradykinin in cats. Circ Res. 1984;55:295–303.
18. Chrissobolis S, Banfi B, Sobey CG, Faraci FM. Role of Nox isoforms in angiotensin II-induced oxidative stress and endothelial dysfunction in brain. J Appl Physiol. 2012;113: 184–91.
19. Chrissobolis S, Faraci FM. The role of oxidative stress and NADPH oxidase in cerebrovascular disease. Trends Mol Med. 2008;14:495–502.
20. Montezano AC, Burger D, Ceravolo GS, Yusuf H, Montero M, Touyz RM. Novel Nox homologues in the vasculature: focusing on Nox4 and Nox5. Clin Sci. 2011;120:131–41.
21. Santhanam AV, d'Uscio LV, Katusic ZS. Erythropoietin increases bioavailability of tetrahydrobiopterin and protects cerebral microvasculature against oxidative stress induced by eNOS uncoupling. J Neurochem. 2014;131:521–9.
22. Santhanam AV, d'Uscio LV, Smith LA, Katusic ZS. Uncoupling of eNOS causes superoxide anion production and impairs NO signaling in the cerebral microvessels of hph-1 mice. J Neurochem. 2012;122:1211–8.
23. Wei EP, Kontos HA, Beckman JS. Mechanisms of cerebral vasodilation by superoxide, hydrogen peroxide, and peroxynitrite. Am J Physiol. 1996;271:H1262–6.
24. Didion SP, Faraci FM. Effects of NADH and NADPH on superoxide levels and cerebral vascular tone. Am J Physiol. 2002;282:H688–95.
25. Park L, Anrather J, Zhou P, Frys K, Wang G, Iadecola C. Exogenous NADPH increases cerebral blood flow through NADPH oxidase-dependent and -independent mechanisms. Arterioscler Thromb Vasc Biol. 2004;24:1860–5.
26. Miller AA, Drummond GR, Sobey CG. Novel isoforms of NADPH-oxidase in cerebral vascular control. Pharmacol Ther. 2006;111:928–48.
27. Cosentino F, Sill JC, Katusic ZS. Role of superoxide anions in the mediation of endothelium-dependent contractions. Hypertension. 1994;23:229–35.
28. Amberg GC, Earley S, Glapa SA. Local regulation of arterial L-type calcium channels by reactive oxygen species. Circ Res. 2010;107:1002–10.
29. Faraci FM. Reactive oxygen species: influence on cerebral vascular tone. J Appl Physiol. 2006;100:739–43.
30. Sobey CG, Heistad DD, Faraci FM. Mechanisms of bradykinin-induced cerebral vasodilatation in rats. Evidence that reactive oxygen species activate K^+ channels. Stroke. 1997;28: 2290–4.
31. Faraci FM, Heistad DD. Regulation of the cerebral circulation: role of endothelium and potassium channels. Physiol Rev. 1998;78:53–97.

32. Faraci FM, Sobey CG. Role of potassium channels in regulation of cerebral vascular tone. J Cereb blood Flow Metabl. 1998;18:1047–63.
33. Faraci FM. Hydrogen peroxide: watery fuel for change in vascular biology. Arterioscler Thromb Vasc Biol. 2006;26:1931–3.
34. Brzezinska AK, Gebremedhin D, Chilian WM, Kalyanaraman B, Elliott SJ. Peroxynitrite reversibly inhibits Ca^{2+}-activated K^{+} channels in rat cerebral artery smooth muscle cells. Am J Physiol. 2000;278:H1883–90.
35. Elliott SJ, Lacey DJ, Chilian WM, Brzezinska AK. Peroxynitrite is a contractile agonist of cerebral artery smooth muscle cells. Am J Physiol. 1998;275:H1585–91.
36. Girouard H, Park L, Anrather J, Zhou P, Iadecola C. Cerebrovascular nitrosative stress mediates neurovascular and endothelial dysfunction induced by angiotensin II. Arterioscler Thromb Vasc Biol. 2007;27:303–9.
37. Maneen MJ, Cipolla MJ. Peroxynitrite diminishes myogenic tone in cerebral arteries: role of nitrotyrosine and F-actin. Am J Physiol. 2007;292:H1042–50.
38. Modrick ML, Didion SP, Sigmund CD, Faraci FM. Role of oxidative stress and AT1 receptors in cerebral vascular dysfunction with aging. Am J Physiol. 2009;296:H1914–9.
39. Didion SP, Kinzenbaw DA, Schrader LI, Faraci FM. Heterozygous CuZn superoxide dismutase deficiency produces a vascular phenotype with aging. Hypertension. 2006;48: 1072–9.
40. Park L, Wang G, Moore J, Girouard H, Zhou P, Anrather J, et al. The key role of transient receptor potential melastatin-2 channels in amyloid-beta-induced neurovascular dysfunction. Nat Comm. 2014;5:5318.
41. Bauer J, Ripperger A, Frantz S, Ergun S, Schwedhelm E, Benndorf RA. Pathophysiology of isoprostanes in the cardiovascular system: implications of isoprostane-mediated thromboxane A_2 receptor activation. Br J Pharmacol. 2014;171:3115–31.
42. Katusic ZS, Austin SA. Endothelial nitric oxide: protector of a healthy mind. Eur Heart J. 2014;35:888–94.
43. Green DJ, Dawson EA, Groenewoud HM, Jones H, Thijssen DH. Is flow-mediated dilation nitric oxide mediated? A meta-analysis. Hypertension. 2014;63:376–82.
44. Lind L, Berglund L, Larsson A, Sundstrom J. Endothelial function in resistance and conduit arteries and 5-year risk of cardiovascular disease. Circulation. 2011;123:1545–51.
45. Flammer AJ, Luscher TF. Three decades of endothelium research: from the detection of nitric oxide to the everyday implementation of endothelial function measurements in cardiovascular diseases. Swiss Med Wkly. 2010;140:w13122.
46. Volpe M, Iaccarino G, Vecchione C, Rizzoni D, Russo R, Rubattu S, et al. Association and cosegregation of stroke with impaired endothelium-dependent vasorelaxation in stroke prone, spontaneously hypertensive rats. J Clin Invest. 1996;98:256–61.
47. Hill-Eubanks DC, Gonzales AL, Sonkusare SK, Nelson MT. Vascular TRP channels: performing under pressure and going with the flow. Physiology. 2014;29:343–60.
48. Kamouchi M, Ago T, Kitazono T. Brain pericytes: emerging concepts and functional roles in brain homeostasis. Cell Mol Neurobiol. 2011;31:175–93.
49. Armulik A, Genove G, Betsholtz C. Pericytes: developmental, physiological, and pathological perspectives, problems, and promises. Dev Cell. 2011;21:193–215.
50. Hall CN, Reynell C, Gesslein B, Hamilton NB, Mishra A, Sutherland BA, et al. Capillary pericytes regulate cerebral blood flow in health and disease. Nature. 2014;508:55–60.
51. De Silva TM, Modrick ML, Ketsawatsomkron P, Lynch C, Chu Y, Pelham CJ, et al. Role of peroxisome proliferator-activated receptor-gamma in vascular muscle in the cerebral circulation. Hypertension. 2014;64:1088–93.
52. Francis SH, Busch JL, Corbin JD, Sibley D. cGMP-dependent protein kinases and cGMP phosphodiesterases in nitric oxide and cGMP action. Pharmacol Rev. 2010;62:525–63.
53. Sobey CG, Faraci FM. Effects of a novel inhibitor of guanylyl cyclase on dilator responses of mouse cerebral arterioles. Stroke. 1997;28:837–42.
54. Katakam PV, Domoki F, Lenti L, Gaspar T, Institoris A, Snipes JA, et al. Cerebrovascular responses to insulin in rats. J Cereb Blood Flow Metab. 2009;29:1955–67.

55. Chrissobolis S, Drummond GR, Faraci FM, Sobey CG. Chronic aldosterone administration causes Nox2-mediated increases in reactive oxygen species production and endothelial dysfunction in the cerebral circulation. J Hypertens. 2014;32:1815–21.
56. De Silva TM, Brait VH, Drummond GR, Sobey CG, Miller AA. Nox2 oxidase activity accounts for the oxidative stress and vasomotor dysfunction in mouse cerebral arteries following ischemic stroke. PLoS One. 2011;6, e28393.
57. Girouard H, Park L, Anrather J, Zhou P, Iadecola C. Angiotensin II attenuates endothelium-dependent responses in the cerebral microcirculation through Nox-2-derived radicals. Arterioscler Thromb Vasc Biol. 2006;26:826–32.
58. Kazama K, Anrather J, Zhou P, Girouard H, Frys K, Milner TA, et al. Angiotensin II impairs neurovascular coupling in neocortex through NADPH oxidase-derived radicals. Circ Res. 2004;95:1019–26.
59. Lynch CM, Kinzenbaw DA, Chen X, Zhan S, Mezzetti E, Filosa J, et al. Nox2-derived superoxide contributes to cerebral vascular dysfunction in diet-induced obesity. Stroke. 2013;44:3195–201.
60. Park L, Anrather J, Girouard H, Zhou P, Iadecola C. Nox2-derived reactive oxygen species mediate neurovascular dysregulation in the aging mouse brain. J Cereb Blood Flow Metab. 2007;27:1908–18.
61. Park L, Anrather J, Zhou P, Frys K, Pitstick R, Younkin S, et al. NADPH-oxidase-derived reactive oxygen species mediate the cerebrovascular dysfunction induced by the amyloid beta peptide. J Neurosci. 2005;25:1769–77.
62. Park L, Zhou P, Pitstick R, Capone C, Anrather J, Norris EH, et al. Nox2-derived radicals contribute to neurovascular and behavioral dysfunction in mice overexpressing the amyloid precursor protein. Proc Natl Acad Sci. 2008;105:1347–52.
63. Walker AE, Henson GD, Reihl KD, Nielson EI, Morgan RG, Lesniewski LA, et al. Beneficial effects of lifelong caloric restriction on endothelial function are greater in conduit arteries compared to cerebral resistance arteries. Age. 2014;36:559–69.
64. Nicolakakis N, Aboulkassim T, Ongali B, Lecrux C, Fernandes P, Rosa-Neto P, et al. Complete rescue of cerebrovascular function in aged Alzheimer's disease transgenic mice by antioxidants and pioglitazone, a peroxisome proliferator-activated receptor gamma agonist. J Neurosci. 2008;28:9287–96.
65. Tong XK, Lecrux C, Rosa-Neto P, Hamel E. Age-dependent rescue by simvastatin of Alzheimer's disease cerebrovascular and memory deficits. J Neurosci. 2012;32:4705–15.
66. Wardlaw JM, Smith C, Dichgans M. Mechanisms of sporadic cerebral small vessel disease: insights from neuroimaging. Lancet Neurol. 2013;12:483–97.
67. Chan SL, Sweet JG, Cipolla MJ. Treatment for cerebral small vessel disease: effect of relaxin on the function and structure of cerebral parenchymal arterioles during hypertension. FASEB J. 2013;27:3917–27.
68. Cipolla MJ, Bullinger LV. Reactivity of brain parenchymal arterioles after ischemia and reperfusion. Microcirculation. 2008;15:495–501.
69. Nakahata K, Kinoshita H, Azma T, Matsuda N, Hama-Tomioka K, Haba M, et al. Propofol restores brain microvascular function impaired by high glucose via the decrease in oxidative stress. Anesthesiology. 2008;108:269–75.
70. Gauthier KM, Campbell WB, McNeish AJ. Regulation of $K_{Ca}2.3$ and endothelium-dependent hyperpolarization (EDH) in the rat middle cerebral artery: the role of lipoxygenase metabolites and isoprostanes. Peer J. 2014;2:414.
71. Mayhan WG. Role of prostaglandin H_2-thromboxane A_2 in responses of cerebral arterioles during chronic hypertension. Am J Physiol. 1992;262:H539–43.
72. Mayhan WG, Faraci FM, Heistad DD. Responses of cerebral arterioles to adenosine 5′-diphosphate, serotonin, and the thromboxane analogue U-46619 during chronic hypertension. Hypertension. 1988;12:556–61.
73. Mayhan WG, Simmons LK, Sharpe GM. Mechanism of impaired responses of cerebral arterioles during diabetes mellitus. Am J Physiol. 1991;260:H319–26.

74. De Caterina R, Libby P, Peng HB, Thannickal VJ, Rajavashisth TB, Gimbrone Jr MA, et al. Nitric oxide decreases cytokine-induced endothelial activation. Nitric oxide selectively reduces endothelial expression of adhesion molecules and proinflammatory cytokines. J Clin Invest. 1995;96:60–8.
75. Shimokawa H, Satoh K. Light and dark of reactive oxygen species for vascular function. J Cardiovasc Pharmacol. 2015;65:412–8.
76. Dong M, Yan BP, Liao JK, Lam YY, Yip GW, Yu CM. Rho-kinase inhibition: a novel therapeutic target for the treatment of cardiovascular diseases. Drug Disc Today. 2010;15:622–9.
77. Sawada N, Liao JK. Rho/Rho-associated coiled-coil forming kinase pathway as therapeutic targets for statins in atherosclerosis. Antioxid Redox Signal. 2014;20:1251–67.
78. Didion SP, Lynch CM, Baumbach GL, Faraci FM. Impaired endothelium-dependent responses and enhanced influence of Rho-kinase in cerebral arterioles in type II diabetes. Stroke. 2005;36:342–7.
79. Iadecola C, Zhang F, Niwa K, Eckman C, Turner SK, Fischer E, et al. SOD1 rescues cerebral endothelial dysfunction in mice overexpressing amyloid precursor protein. Nat Neurosci. 1999;2:157–61.
80. Zhang L, Papadopoulos P, Hamel E. Endothelial TRPV4 channels mediate dilation of cerebral arteries: impairment and recovery in cerebrovascular pathologies related to Alzheimer's disease. Br J Pharmacol. 2013;170:661–70.
81. Faraci F. Cerebral vascular dysfunction with aging. In: Masoro EJ, Austad S, editors. Handbook of the biology of aging. 7th ed. New York, NY: Academic; 2011. p. 405–18.
82. Katusic ZS, Marshall JJ, Kontos HA, Vanhoutte PM. Similar responsiveness of smooth muscle of the canine basilar artery to EDRF and nitric oxide. Am J Physiol. 1989;257:H1235–9.
83. Kontos HA, Wei EP, Marshall JJ. In vivo bioassay of endothelium-derived relaxing factor. Am J Physiol. 1988;255:H1259–62.
84. Nelson CW, Wei EP, Povlishock JT, Kontos HA, Moskowitz MA. Oxygen radicals in cerebral ischemia. Am J Physiol. 1992;263:H1356–62.
85. Modrick ML, Didion SP, Lynch CM, Dayal S, Lentz SR, Faraci FM. Role of hydrogen peroxide and the impact of glutathione peroxidase-1 in regulation of cerebral vascular tone. J Cereb Blood Flow Metab. 2009;29:1130–7.
86. Bryan Jr RM, You J, Golding EM, Marrelli SP. Endothelium-derived hyperpolarizing factor: a cousin to nitric oxide and prostacyclin. Anesthesiology. 2005;102:1261–77.
87. Dunn KM, Nelson MT. Neurovascular signaling in the brain and the pathological consequences of hypertension. Am J Physiol. 2014;306:H1–14.
88. Iadecola C. The pathobiology of vascular dementia. Neuron. 2013;80:844–66.
89. Chen BR, Kozberg MG, Bouchard MB, Shaik MA, Hillman EM. A critical role for the vascular endothelium in functional neurovascular coupling in the brain. J Am Heart Assoc. 2014;3:e000787.
90. Faraci FM, Heistad DD. Regulation of large cerebral arteries and cerebral microvascular pressure. Circ Res. 1990;66:8–17.
91. Fujii K, Heistad DD, Faraci FM. Flow-mediated dilatation of the basilar artery in vivo. Circ Res. 1991;69:697–705.
92. Joutel A, Faraci FM. Cerebral small vessel disease: insights and opportunities from mouse models of collagen IV-related small vessel disease and cerebral autosomal dominant arteriopathy with subcortical infarcts and leukoencephalopathy. Stroke. 2014;45:1215–21.
93. Vetri F, Xu H, Paisansathan C, Pelligrino DA. Impairment of neurovascular coupling in type 1 diabetes mellitus in rats is linked to PKC modulation of BK_{Ca} and K_{ir} channels. Am J Physiol. 2012;302:H1274–84.
94. Kazama K, Wang G, Frys K, Anrather J, Iadecola C. Angiotensin II attenuates functional hyperemia in the mouse somatosensory cortex. Am J Physiol. 2003;285:H1890–9.
95. Jackman K, Iadecola C. Neurovascular regulation in the ischemic brain. Antioxid Redox Signal. 2015;22:149–60.
96. Park L, Koizumi K, El Jamal S, Zhou P, Previti ML, Van Nostrand WE, et al. Age-dependent neurovascular dysfunction and damage in a mouse model of cerebral amyloid angiopathy. Stroke. 2014;45:1815–21.

97. Dabertrand F, Hannah RM, Pearson JM, Hill-Eubanks DC, Brayden JE, Nelson MT. Prostaglandin E_2, a postulated astrocyte-derived neurovascular coupling agent, constricts rather than dilates parenchymal arterioles. J Cereb Blood Flow Metab. 2013;33:479–82.
98. Shi Y, Liu X, Gebremedhin D, Falck JR, Harder DR, Koehler RC. Interaction of mechanisms involving epoxyeicosatrienoic acids, adenosine receptors, and metabotropic glutamate receptors in neurovascular coupling in rat whisker barrel cortex. J Cereb Blood Flow Metab. 2008;28:111–25.
99. Gebremedhin D, Weinberger B, Lourim D, Harder DR. Adenosine can mediate its actions through generation of reactive oxygen species. J Cereb Blood Flow Metab. 2010;30: 1777–90.
100. Brian Jr JE. Carbon dioxide and the cerebral circulation. Anesthesiology. 1998;88:1365–86.
101. Kontos HA, Wei EP, Raper AJ, Patterson Jr JL. Local mechanism of CO_2 action of cat pial arterioles. Stroke. 1977;8:226–9.
102. Niwa K, Haensel C, Ross ME, Iadecola C. Cyclooxygenase-1 participates in selected vasodilator responses of the cerebral circulation. Circ Res. 2001;88:600–8.
103. Leffler CW, Mirro R, Thompson C, Shibata M, Armstead WM, Pourcyrous M, et al. Activated oxygen species do not mediate hypercapnia-induced cerebral vasodilation in newborn pigs. Am J Physiol. 1991;261:H335–42.
104. Zhang F, Slungaard A, Vercellotti GM, Iadecola C. Superoxide-dependent cerebrovascular effects of homocysteine. Am J Physiol. 1998;274:R1704–11.
105. Niwa K, Carlson GA, Iadecola C. Exogenous A beta1-40 reproduces cerebrovascular alterations resulting from amyloid precursor protein overexpression in mice. J Cereb Blood Flow Metab. 2000;20:1659–68.
106. Cipolla MJ. The cerebral circulation. Integrated systems physiology: from molecule to function. San Rafael, CA: Morgan & Claypool Life Sciences; 2009. p. 1–59.
107. Faraci FM, Baumbach GL, Heistad DD. Myogenic mechanisms in the cerebral circulation. J Hypertens. 1989;7:S61–4.
108. Kontos HA, Wei EP, Dietrich WD, Navari RM, Povlishock JT, Ghatak NR, et al. Mechanism of cerebral arteriolar abnormalities after acute hypertension. Am J Physiol. 1981;240: H511–27.
109. Wei EP, Kontos HA, Dietrich WD, Povlishock JT, Ellis EF. Inhibition by free radical scavengers and by cyclooxygenase inhibitors of pial arteriolar abnormalities from concussive brain injury in cats. Circ Res. 1981;48:95–103.
110. Gebremedhin D, Terashvili M, Wickramasekera N, Zhang DX, Rau N, Miura H, et al. Redox signaling via oxidative inactivation of PTEN modulates pressure-dependent myogenic tone in rat middle cerebral arteries. PLoS One. 2013;8, e68498.
111. Lim M, Choi SK, Cho YE, Yeon SI, Kim EC, Ahn DS, et al. The role of sphingosine kinase 1/sphingosine-1-phosphate pathway in the myogenic tone of posterior cerebral arteries. PLoS One. 2012;7, e35177.
112. Butcher JT, Goodwill AG, Stanley SC, Frisbee JC. Differential impact of dilator stimuli on increased myogenic activation of cerebral and skeletal muscle resistance arterioles in obese zucker rats. Microcirculation. 2013;20:579–89.
113. Phillips SA, Sylvester FA, Frisbee JC. Oxidant stress and constrictor reactivity impair cerebral artery dilation in obese Zucker rats. Am J Physiol. 2005;288:R522–30.
114. Didion SP, Ryan MJ, Didion LA, Fegan PE, Sigmund CD, Faraci FM. Increased superoxide and vascular dysfunction in CuZnSOD-deficient mice. Circ Res. 2002;91:938–44.
115. Faraci FM, Modrick ML, Lynch CM, Didion LA, Fegan PE, Didion SP. Selective cerebral vascular dysfunction in Mn-SOD-deficient mice. J Appl Physiol. 2006;100:2089–93.
116. Kitayama J, Yi C, Faraci FM, Heistad DD. Modulation of dilator responses of cerebral arterioles by extracellular superoxide dismutase. Stroke. 2006;37:2802–6.
117. Brown KA, Didion SP, Andresen JJ, Faraci FM. Effect of aging, MnSOD deficiency, and genetic background on endothelial function: evidence for MnSOD haploinsufficiency. Arterioscler Thromb Vasc Biol. 2007;27:1941–6.

118. Chrissobolis S, Faraci FM. Sex differences in protection against angiotensin II-induced endothelial dysfunction by manganese superoxide dismutase in the cerebral circulation. Hypertension. 2010;55:905–10.
119. Didion SP, Kinzenbaw DA, Faraci FM. Critical role for CuZn-superoxide dismutase in preventing angiotensin II-induced endothelial dysfunction. Hypertension. 2005;46:1147–53.
120. Chrissobolis S, Didion SP, Kinzenbaw DA, Schrader LI, Dayal S, Lentz SR, et al. Glutathione peroxidase-1 plays a major role in protecting against angiotensin II-induced vascular dysfunction. Hypertension. 2008;51:872–7.
121. Ketsawatsomkron P, Pelham CJ, Groh S, Keen HL, Faraci FM, Sigmund CD. Does peroxisome proliferator-activated receptor-gamma protect from hypertension directly through effects in the vasculature? J Biol Chem. 2010;285:9311–6.
122. Marchesi C, Schiffrin EL. Peroxisome proliferator-activated receptors and the vascular system: beyond their metabolic effects. J Am Soc Hypertens. 2008;2:227–38.
123. Plutzky J. The PPAR-RXR transcriptional complex in the vasculature: energy in the balance. Circ Res. 2011;108:1002–16.
124. Lincoff AM, Wolski K, Nicholls SJ, Nissen SE. Pioglitazone and risk of cardiovascular events in patients with type 2 diabetes mellitus: a meta-analysis of randomized trials. JAMA. 2007;298:1180–8.
125. Ryan MJ, Didion SP, Mathur S, Faraci FM, Sigmund CD. PPAR gamma agonist rosiglitazone improves vascular function and lowers blood pressure in hypertensive transgenic mice. Hypertension. 2004;43:661–6.
126. Cipolla MJ, Bishop N, Vinke RS, Godfrey JA. PPAR gamma activation prevents hypertensive remodeling of cerebral arteries and improves vascular function in female rats. Stroke. 2010;41:1266–70.
127. Barroso I, Gurnell M, Crowley VE, Agostini M, Schwabe JW, Soos MA, et al. Dominant negative mutations in human PPAR gamma associated with severe insulin resistance, diabetes mellitus and hypertension. Nature. 1999;402:880–3.
128. Keen HL, Halabi CM, Beyer AM, de Lange WJ, Liu X, Maeda N, et al. Bioinformatic analysis of gene sets regulated by ligand-activated and dominant-negative peroxisome proliferator-activated receptor gamma in mouse aorta. Arterioscler Thromb Vasc Biol. 2010;30:518–25.
129. Beyer AM, Baumbach GL, Halabi CM, Modrick ML, Lynch CM, Gerhold TD, et al. Interference with PPAR gamma signaling causes cerebral vascular dysfunction, hypertrophy, and remodeling. Hypertension. 2008;51:867–71.
130. Beyer AM, de Lange WJ, Halabi CM, Modrick ML, Keen HL, Faraci FM, et al. Endothelium-specific interference with peroxisome proliferator activated receptor gamma causes cerebral vascular dysfunction in response to a high-fat diet. Circ Res. 2008;103:654–61.
131. Lyle AN, Griendling KK. Modulation of vascular smooth muscle signaling by reactive oxygen species. Physiology. 2006;21:269–80.
132. Reckelhoff JF, Romero JC. Role of oxidative stress in angiotensin-induced hypertension. Am J Physiol. 2003;284:R893–912.
133. Balakumar P, Jagadeesh G. A century old renin-angiotensin system still grows with endless possibilities: a T receptor signaling cascades in cardiovascular physiopathology. Cell Signal. 2014;26:2147–60.
134. Santos RA. Angiotensin-(1–7). Hypertension. 2014;63:1138–47.
135. Pena Silva RA, Chu Y, Miller JD, Mitchell IJ, Penninger JM, Faraci FM, et al. Impact of ACE2 deficiency and oxidative stress on cerebrovascular function with aging. Stroke. 2012;43:3358–63.
136. Higashi Y, Maruhashi T, Noma K, Kihara Y. Oxidative stress and endothelial dysfunction: clinical evidence and therapeutic implications. Trends Cardiovasc Med. 2014;24:165–9.
137. Rodriguez-Manas L, El-Assar M, Vallejo S, Lopez-Doriga P, Solis J, Petidier R, et al. Endothelial dysfunction in aged humans is related with oxidative stress and vascular inflammation. Aging Cell. 2009;8:226–38.
138. Wray DW, Nishiyama SK, Harris RA, Zhao J, McDaniel J, Fjeldstad AS, et al. Acute reversal of endothelial dysfunction in the elderly after antioxidant consumption. Hypertension. 2012;59:818–24.

139. Heart Protection Study Collaborative G. MRC/BHF Heart Protection Study of antioxidant vitamin supplementation in 20,536 high-risk individuals: a randomised placebo-controlled trial. Lancet. 2002;360:23–33.
140. Yamada M, Lamping KG, Duttaroy A, Zhang W, Cui Y, Bymaster FP, et al. Cholinergic dilation of cerebral blood vessels is abolished in M_5 muscarinic acetylcholine receptor knockout mice. Proc Natl Acad Sci. 2001;98:14096–101.
141. Sullivan MN, Earley S. TRP channel Ca^{2+} sparklets: fundamental signals underlying endothelium-dependent hyperpolarization. Am J Physiol. 2013;305:C999–1008.
142. Ngai AC, Winn HR. Modulation of cerebral arteriolar diameter by intraluminal flow and pressure. Circ Res. 1995;77:832–40.
143. Ago T, Kitazono T, Kuroda J, Kumai Y, Kamouchi M, Ooboshi H, et al. NAD(P)H oxidases in rat basilar arterial endothelial cells. Stroke. 2005;36:1040–6.
144. Paravicini TM, Chrissobolis S, Drummond GR, Sobey CG. Increased NADPH-oxidase activity and Nox4 expression during chronic hypertension is associated with enhanced cerebral vasodilatation to NADPH in vivo. Stroke. 2004;35:584–9.
145. Fang Q, Sun H, Arrick DM, Mayhan WG. Inhibition of NADPH oxidase improves impaired reactivity of pial arterioles during chronic exposure to nicotine. J Appl Physiol. 2006;100: 631–6.
146. Mayhan WG, Arrick DM, Sharpe GM, Patel KP, Sun H. Inhibition of NAD(P)H oxidase alleviates impaired NOS-dependent responses of pial arterioles in type 1 diabetes mellitus. Microcirculation. 2006;13:567–75.
147. Miller AA, Drummond GR, Mast AE, Schmidt HH, Sobey CG. Effect of gender on NADPH-oxidase activity, expression, and function in the cerebral circulation: role of estrogen. Stroke. 2007;38:2142–9.
148. De Silva TM, Broughton BR, Drummond GR, Sobey CG, Miller AA. Gender influences cerebral vascular responses to angiotensin II through Nox2-derived reactive oxygen species. Stroke. 2009;40:1091–7.
149. Miller AA, De Silva TM, Judkins CP, Diep H, Drummond GR, Sobey CG. Augmented superoxide production by Nox2-containing NADPH oxidase causes cerebral artery dysfunction during hypercholesterolemia. Stroke. 2010;41:784–9.
150. Kleinschnitz C, Grund H, Wingler K, Armitage ME, Jones E, Mittal M, et al. Post-stroke inhibition of induced NADPH oxidase type 4 prevents oxidative stress and neurodegeneration. PLoS Biol. 2010;8.
151. Akopov SE, Grigorian MR, Gabrielian ES. The endothelium-dependent relaxation of human middle cerebral artery: effects of activated neutrophils. Experientia. 1992;48:34–6.
152. Wei EP, Kontos HA, Christman CW, DeWitt DS, Povlishock JT. Superoxide generation and reversal of acetylcholine-induced cerebral arteriolar dilation after acute hypertension. Circ Res. 1985;57:781–7.
153. Mayhan WG, Arrick DM, Sharpe GM, Sun H. Age-related alterations in reactivity of cerebral arterioles: role of oxidative stress. Microcirculation. 2008;15:225–36.
154. Sun H, Mayhan WG. Temporal effect of alcohol consumption on reactivity of pial arterioles: role of oxygen radicals. Am J Physiol. 2001;280:H992–1001.
155. Sun H, Zheng H, Molacek E, Fang Q, Patel KP, Mayhan WG. Role of NAD(P)H oxidase in alcohol-induced impairment of endothelial nitric oxide synthase-dependent dilation of cerebral arterioles. Stroke. 2006;37:495–500.
156. Sun H, Mayhan WG. Superoxide dismutase ameliorates impaired nitric oxide synthase-dependent dilatation of the basilar artery during chronic alcohol consumption. Brain Res. 2001;891:116–22.
157. Tong XK, Nicolakakis N, Kocharyan A, Hamel E. Vascular remodeling versus amyloid beta-induced oxidative stress in the cerebrovascular dysfunctions associated with Alzheimer's disease. J Neurosci. 2005;25:11165–74.
158. Capone C, Faraco G, Anrather J, Zhou P, Iadecola C. Cyclooxygenase 1-derived prostaglandin E_2 and EP1 receptors are required for the cerebrovascular dysfunction induced by angiotensin II. Hypertension. 2010;55:911–7.

159. Capone C, Faraco G, Park L, Cao X, Davisson RL, Iadecola C. The cerebrovascular dysfunction induced by slow pressor doses of angiotensin-II precedes the development of hypertension. Am J Physiol. 2011;300:H397–407.
160. Johnson AW, Kinzenbaw DA, Modrick ML, Faraci FM. Small-molecule inhibitors of signal transducer and activator of transcription 3 protect against angiotensin II-induced vascular dysfunction and hypertension. Hypertension. 2013;61:437–42.
161. Faraci FM, Lamping KG, Modrick ML, Ryan MJ, Sigmund CD, Didion SP. Cerebral vascular effects of angiotensin II: new insights from genetic models. J Cereb Blood Flow Metab. 2006;26:449–55.
162. Gibson CC, Zhu W, Davis CT, Bowman-Kirigin JA, Chan AC, Ling J, et al. Strategy for identifying repurposed drugs for the treatment of cerebral cavernous malformation. Circulation. 2015;131(3):289–99.
163. Faraco G, Wijasa TS, Park L, Moore J, Anrather J, Iadecola C. Water deprivation induces neurovascular and cognitive dysfunction through vasopressin-induced oxidative stress. J Cereb Blood Flow Metab. 2014;34:852–60.
164. Mayhan WG. Superoxide dismutase partially restores impaired dilatation of the basilar artery during diabetes mellitus. Brain Res. 1997;760:204–9.
165. Didion SP, Lynch CM, Faraci FM. Cerebral vascular dysfunction in TallyHo mice: a new model of type II diabetes. Am J Physiol. 2007;292:H1579–83.
166. Matsumoto T, Kobayashi T, Wachi H, Seyama Y, Kamata K. Vascular NAD(P)H oxidase mediates endothelial dysfunction in basilar arteries from Otsuka Long-Evans Tokushima Fatty (OLETF) rats. Atherosclerosis. 2007;192:15–24.
167. Erdos B, Snipes JA, Miller AW, Busija DW. Cerebrovascular dysfunction in Zucker obese rats is mediated by oxidative stress and protein kinase C. Diabetes. 2004;53:1352–9.
168. De Silva TM, Lynch CM, Grobe JL, Faraci FM. Activation of the central renin angiotensin system (RAS) causes selective cerebrovascular dysfunction (Abstract). FASEB J. 2015;29:646–4.
169. Kontos HA, Wei EP. Endothelium-dependent responses after experimental brain injury. J Neurotrauma. 1992;9:349–54.
170. Kitayama J, Faraci FM, Lentz SR, Heistad DD. Cerebral vascular dysfunction during hypercholesterolemia. Stroke. 2007;38:2136–41.
171. Dayal S, Arning E, Bottiglieri T, Boger RH, Sigmund CD, Faraci FM, et al. Cerebral vascular dysfunction mediated by superoxide in hyperhomocysteinemic mice. Stroke. 2004;35: 1957–62.
172. Xie H, Ray PE, Short BL. NF-kappaB activation plays a role in superoxide-mediated cerebral endothelial dysfunction after hypoxia/reoxygenation. Stroke. 2005;36:1047–52.
173. Capone C, Faraco G, Coleman C, Young CN, Pickel VM, Anrather J, et al. Endothelin 1-dependent neurovascular dysfunction in chronic intermittent hypoxia. Hypertension. 2012; 60:106–13.
174. Kunz A, Park L, Abe T, Gallo EF, Anrather J, Zhou P, et al. Neurovascular protection by ischemic tolerance: role of nitric oxide and reactive oxygen species. J Neurosci. 2007;27: 7083–93.
175. Hernanz R, Briones AM, Alonso MJ, Vila E, Salaices M. Hypertension alters role of iNOS, COX-2, and oxidative stress in bradykinin relaxation impairment after LPS in rat cerebral arteries. Am J Physiol. 2004;287:H225–34.
176. Mayhan WG, Arrick DM, Sharpe GM, Sun H. Nitric oxide synthase-dependent responses of the basilar artery during acute infusion of nicotine. Nicotine Tob Res. 2009;11:270–7.
177. Zhang R, Bai YG, Lin LJ, Bao JX, Zhang YY, Tang H, et al. Blockade of AT1 receptor partially restores vasoreactivity, NOS expression, and superoxide levels in cerebral and carotid arteries of hindlimb unweighting rats. J Appl Physiol. 2009;106:251–8.

Chapter 7
Therapeutic Strategies Harnessing Oxidative Stress to Treat Stroke

Gina Hadley, Ain A. Neuhaus, and Alastair M. Buchan

Abbreviations

AMPA	α-Amino-3-hydroxy-5-methyl-4-isoxazolepropionic acid
ASA	Acetylsalicylic acid
AT-1	Angiotensin-1 receptor 1
AVASAS	Aspirin versus ascorbic acid plus aspirin in stroke
DNA	Deoxyribonucleic acid
EAAT-1	Excitatory amino acid transporter-1
HMG-CoA	3-Hydroxy-3-methyl-glutaryl-CoA reductase
$HO^{\bullet}$	Hydroxyl radical
H_2O_2	Hydrogen peroxide
mRS	Modified Rankin Scale
NADPH	Nicotinamide adenine dinucleotide phosphate
NDMA	*N*-Methyl-D-aspartate
NIHSS	National Institutes of Health Stroke Scale
NO	Nitric oxide
NOS	Nitric oxide synthase
RNS	Reactive nitrogen species

G. Hadley, M.R.C.P. • A.A. Neuhaus, B.A. • A.M. Buchan, F.Med.Sci. (✉)
Acute Stroke Programme, Radcliffe Department of Medicine, John Radcliffe Hospital, University of Oxford, Oxford OX3 9DU, UK
e-mail: gina.hadley@hmc.ox.ac.uk; ain.neuhaus@rdm.ox.ac.uk; alastair.buchan@medsci.ox.ac.uk

M. Rodriguez-Porcel et al. (eds.), *Studies on Atherosclerosis*, Oxidative Stress in Applied Basic Research and Clinical Practice,
DOI 10.1007/978-1-4899-7693-2_7

ROS	Reactive oxygen species
rtPA	Recombinant tissue plasminogen activator
SOD	Superoxide dismutase
VISTA	Virtual International Stroke Trials Archive

Introduction

Stroke is the western world's greatest cause of morbidity and the third biggest cause of mortality [1]. The majority of strokes are ischemic in nature (85 %) and the remainder (15 %) are haemorrhagic. Stroke is a medical emergency and should be viewed as a 'brain attack'. Ischemic stroke is due to the occlusion of blood vessels in the brain. The immediately downstream tissue, the core of the infarct, suffers extensive tissue damage, whereas the surrounding area of reduced cerebral blood flow and metabolism, the penumbra, may be salvageable following reperfusion. Indeed, with the noteworthy exception of tissue plasminogen activator [2] for achieving reperfusion, recent work on endovascular thrombectomy [3] and antiplatelet agents for preventing further coagulation, there is little that has been translated into effective treatment for stroke patients.

Ischemia causes uncontrolled depolarisation of neurons, release of glutamate and subsequent stimulation of *N*-methyl-D-aspartate (NDMA), α-Amino-3-hydroxy-5-methyl-4-isoxazolepropionic acid (AMPA) and metabotropic receptors, resulting in an influx of calcium into the cell. In neurons, this causes a specific form of nitric oxide synthase (NOS) to form nitric oxide (NO), which in excess leads to the formation of free radicals that interfere with the electron transport chain in mitochondria by binding to cytochromes and preventing oxidative phosphorylation [4].

The concept of free radicals inducing damage in stroke is not new [5, 6]. The brain is especially vulnerable to oxidative stress as it has a relatively low endogenous antioxidant capacity, is a high consumer of oxygen and contains high amounts of iron and unsaturated lipids [6]. Oxidative stress is a major mechanism responsible for damage in ischemic stroke via the production of reactive oxygen species (ROS; e.g. the hydroxyl radical ($HO^•$), hydrogen peroxide (H_2O_2), and superoxide anion radical ($O_2^{•}-$)) and reactive nitrogen species (RNS), occurring as a direct result of the insult as well as during reperfusion. Examples of enzymes able to form ROS are uncoupled NOS, xanthine oxidase and members of the cytochrome P450 family. They all require an initial oxidation step and have other functions. Nicotinamide adenine dinucleotide phosphate (NADPH) oxidases on the other hand only produce ROS [7]. The action of phospholipase A2 and cyclooxygenase generates some of the free radicals responsible for lipid peroxidation [8].

Reperfusion itself is a well-known cause of the production of ROS [9]. Determining the clinical relevance of pathophysiological mechanisms based on animal models is not necessarily straightforward due to confounding factors associated with induction of the stroke, nature of the ischemia and the model organism in question (reviewed by [10]). This also applies to free radicals, particularly due to

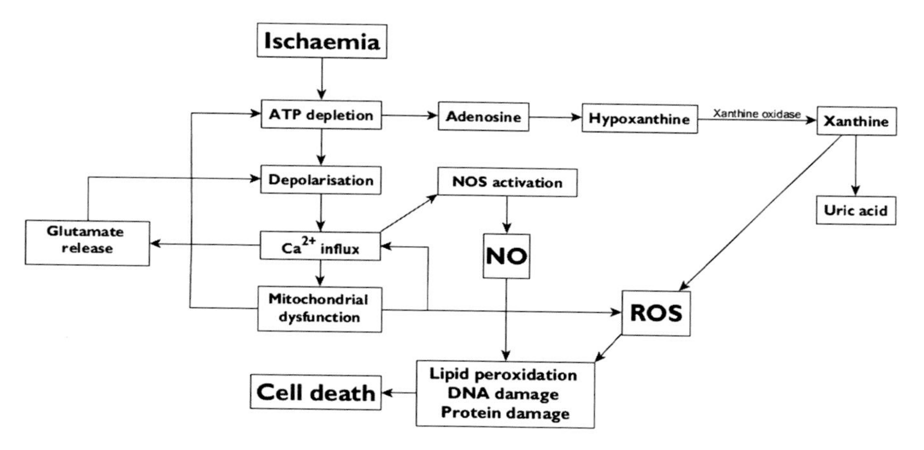

Fig. 7.1 The ischemic cascade. A summary of the pathways discussed in the chapter. Adapted from [15, 16]

their role in mediating re-oxygenation damage [11]. It has been suggested that transient models of ischemia, where cerebral arteries are mechanically occluded for a period of time and then reopened do not accurately reflect the clinical setting, as reperfusion even in the case of thrombolysis is protracted, whereas these models cause a sharp return to baseline or near-baseline blood flow. As a consequence, the molecular cascades leading to cell death may fundamentally differ, including the involvement of free radicals [12]. However, even in the case of permanent ischemia—highly representative of most strokes—free radical scavengers exhibit protective effects [13], supporting their validity as a therapeutic target.

There are three major ways in which the cell copes with oxidative stress (reviewed by [14]). Firstly, by the regulation of *de novo* (ROS) generation; second, by the presence of ROS scavenger molecules and neutralising enzymes e.g. superoxide dismutase (SOD) and hemeoxygenase; third via cell repair mechanisms targeted at damaged proteins, lipids and DNA. Once these mechanisms become saturated, the balance is tipped towards cell death via apoptosis, necrosis and autophagy. Pharmacologically, the majority of interventions targeting free radicals have been aimed at inhibition of enzymes producing ROS/RNS, and scavenging the radicals that are already present. See Fig. 7.1 for a summary of the ischemic cascade discussed in this chapter.

Allopurinol and Uric Acid

Under ischemic conditions, xanthine oxidase is produced from xanthine dehydrogenase via proteolytic modification, leading to increased superoxide anion production. However, uric acid, another product of purine metabolism via xanthine oxidase, is itself a major endogenous antioxidant [17].

Uric Acid

Uric acid scavenges hydroxyl radicals, hydrogen peroxide, and peroxynitrite. It also chelates transition metals, inhibits lipid peroxidation and the Fenton reaction which produces the hydroxyl radical [18]. In addition it may prevent the degradation of superoxide dismutase permitting the removal of superoxide. It has also been shown to act on astroglia in order to upregulate excitatory amino acid transporter-1 (EAAT-1), a redox-sensitive glutamate transporter, to mitigate the effects of glutamate toxicity in spinal cord neurons [19].

There is controversy as to whether or not a high serum urate is protective or detrimental in the context of ischemic stroke. Patients with increased levels of serum uric acid are at increased risk of myocardial infarction, cerebral infarction and congestive heart failure, demonstrated by the Apolipoprotein MOrtality RISk study (AMORIS) which included 417,734 men and women in Norway [20]. A systematic review and meta-analysis consisting of 16 studies (238,449 adults) also showed that hyperuricemia is associated with modestly increased stroke incidence and mortality [21]. In a prospective cohort study of 41,879 men and 48,514 women in Taiwan, hyperuricemia was an independent risk factor of mortality from all causes including total cerebrovascular disease and ischemic stroke [22]. This does not, however, preclude a benefit if administered when ischemia has occurred. Indeed, once confounders are controlled for in many studies, including Framingham, the relationship between raised uric acid and an increase in cardiovascular events is not significant [23, 24]. What remains elusive is a causal link [25].

In the past, uric acid has been seen as a risk factor for the development of cerebral and myocardial ischemia, notably in the Rotterdam Study [26]. In Norway, a population-based prospective cohort study was undertaken (2696 men and 3004 women), serum uric acid was linked with a 31 % increased risk of stroke in men [27].

In one retrospective study, serum urate was measured in 2498/3731 patients with ischemic stroke (90 %) or primary intracranial haemorrhage (10 %). Higher levels predicted a lower chance of a good 90-day outcome and higher rate of vascular events [28]. In addition, using data from 852 patients from the Virtual International Stroke Trials Archive (VISTA), some of the same investigators found that serum urate level was not independently associated with poor 90-day outcome (measured by the modified Rankin Scale (mRS)). However, following univariate analysis an association was found (OR 1.57, 95 % CI 1.02–2.42) [29, 30]. In a small Polish study bilirubin and uric acid were found to be poor prognostic factors in ischemic stroke [31]. In a Mexican study of 463 patients, low serum uric acid correlated with a good short term outcome; however, it was stated that serum uric acid is a marker of the size of the stroke rather than an independent predictor of patient outcome [32].

From an observational study, stroke patients with higher serum levels of uric acid on admission had better outcomes at follow up [33]. In a study of 585 young Chinese adults with acute ischemic stroke, serum uric acid was demonstrated to be an independent predictor for good clinical outcome measured by the mRS [34].

In 218 consecutive patients receiving intravenous thrombolysis, serum urate was measured with outcomes determined using National Institutes of Health Stroke Scale (NIHSS) 24 h after onset and the mRS after 3 months. Higher serum urate levels may be associated with better outcomes but depended on gender and initial stroke severity [35].

The recently published URICO-ICTUS trial did not show any benefit compared to the placebo group [36]. The phase 2b/3 multicentre trial, interventional, randomised, double-blind and vehicle-controlled efficacy study was conducted between 2011 and 2013, 411/421 of patients recruited were included in the study. Patients in treatment arm ($n=211$) received an intravenous infusion of 1 g of uric acid in a vehicle containing 0.1 % lithium carbonate and 5 % mannitol over 90 min. The control arm ($n=200$) received vehicle alone. The primary outcome was the proportion of patients deemed excellent using the mRS.

Preclinical studies had, however, suggested that there was an additive effect when uric acid and recombinant tissue plasminogen activator (rtPA) were given at the same time [37]. In the proof-of-concept study preceding the main trial, it had been found that uric acid infusion avoided early decay in serum uric acid levels and decreased lipid peroxidation marker malondialdehyde [38, 39].

Allopurinol

Xanthine oxidase is also a prime example of the complexity and intertwined nature of free radical cascades: in addition to its role in uric acid production, xanthine oxidase may have detrimental effects as a consequence of superoxide production. Allopurinol is a xanthine oxidase inhibitor that has been tested in several clinical trials. In a small randomised, double-blind, controlled study with 35 of 40 patients completing the protocol, the effect of a 3-month course of 300 mg allopurinol once daily was compared to placebo [29, 30]. Cerebrovascular reactivity (CVR) was used as the outcome measure in subcortical stroke and no significance was found. Clinical trials usually look at functional outcome using mRS or NIHSS which would be more meaningful in the clinical setting. A further small Iranian study (70 patients) found that allopurinol improved the functional status of patients at 3 months (using mRS). They found no association between final serum uric acid level and serum uric acid depletion and final favourable functional status [40].

An explanation for allopurinol's protective effect could be due to an alternative mechanism of action such as its effects on the vasculature. For example, allopurinol reduced arterial wave reflection in stroke survivors, suggesting improved vascular function [41]. Alternatively, any putative benefit might be through an effect on inflammatory pathways. In a small (50 patients) randomised, double-blind, placebo-controlled trial allopurinol was shown to reduce the rise in intercellular adhesion molecule-1 levels following stroke [42], suggesting a diminished inflammatory change in endothelial cells.

The possibility has already been raised that potential therapy will eventually involve both urate and allopurinol [43].

Edaravone

Edaravone, a low-molecular weight, lipophilic compound, readily crosses the blood brain barrier and is a scavenger of hydroxyl, peroxyl, and superoxide radicals. It was approved in 2001 in Japan for use in ischemic stroke within 24 h of the attack ([44], Amaro and Chamorro 2011). The neuroprotective effects of edaravone are not thought to be solely due to its scavenging ability, but also through inhibition of lipoxygenase and the oxidation of low-density lipoprotein in addition to combating microglia-induced neurotoxicity, suppressing delayed neuronal death and reducing inflammation [44].

There have been several clinical trials to evaluate edaravone after stroke. In a multicentre, randomised, placebo-controlled, double-blind study of 250 patients (two excluded due to haemorrhage), with the majority of strokes (approximately 80 %) being thrombotic and the remainder embolic, patients were treated within a 72 h window and edaravone was infused twice daily for 14 days at a dose of 30 mg. The mRS was used to assess outcome, the most apparent improvement was at within 24 h of onset; however, all groups up to a treatment window of 72 h demonstrated clinical improvement [45].

In a randomised placebo-controlled trial of edaravone administered twice daily over 14 days, 72 % of patients in the edaravone group compared with 40 % of the control had favourable 90-day Barthel index scores ($p<0.005$). This suggested that edaravone improved functional outcome [46].

Unno et al. [47] attempted to correlate the amount of edaravone used with functional outcomes in other studies demonstrating using the Functional Independence Measure-Motor (DeltaFIM-M) or Barthel Index (DeltaBI) score that there was a dose-dependent effect of 'rehabilitation gain' [47]. Dose ranges were divided into 'tertiles' in a review of edaravone; it was pointed out that the highest dose in this study is actually less than that used in previous studies [44].

In a Cochrane review of edaravone for acute ischemic stroke, three trials were combined for analysis with a total of 496 patients [45, 48, 49]. Trials containing patients with acute ischemic stroke that were randomised comparing edaravone with placebo or no intervention were included. All studies used a dose of 60 mg per day and duration of treatment was 14 days. Taking into account small studies and a moderate risk of bias, edaravone appeared to significantly increase the amount of patients with neurological improvement (risk ratio (RR) 1.99, 95 % confidence interval (CI) 1.60–2.49) [50]. In a meta-analysis of randomised controlled trials edaravone improved neurological impairment related to ischemic stroke and intracranial haemorrhage, but more evidence was needed for a decrease in death or long-term disability [51].

In a retrospective study of 625 consecutive patients (331 males and 294 females), 237 received both edaravone and conventional treatment and the remainder (388) received conventional treatment. The study was carried out in ischemic stroke patients within 48 h of stroke onset. Clinical outcomes were measured using the mRS. Although treatment with edaravone did not reach significance, there was a tendency towards benefit [52].

There have also been studies looking at lacunar infarction. This type of infarct has a propensity to occur in the white matter. White matter has considerable amounts of oligodendrocytes and myelin which have high levels of lipids and iron and are therefore prone to free radical damage [53]. A small study ($n=70$) of patients with lacunar stroke suggested that patients with mild strokes (NIHSS ≤ 8) benefit from edaravone. There was a significant benefit in mRS scores for patients taking edaravone ($p=0.035$). This was not correlated with time to treatment. Final analysis, however, only included 21 patients in the edaravone group compared to 49 patients in the control group [44, 54]. A further small study retrospectively looked at lacunar stroke patients, 124 consecutive patients (59 edaravone and conventional therapy, 65 conventional therapy alone) [55]. There was no difference between the groups when considering NIHSS at admission or discharge, but the edaravone group had a greater reduction in motor palsy score compared to control group ($p=0.006$) ([55] as reviewed in [44]).

In a further historical-controlled cohort of similar patients with cardioembolic stroke (edaravone for 7 days, $n=141$ vs control, $n=114$), edaravone was shown only to be effective in patients with NIHSS ≤ 7 [56]. In a trial looking at internal carotid artery occlusion, 30 patients with a baseline NIHSS >15 were given intravenous edaravone for 14 days and compared with a control group consisting of 31 similar patients. Glycerol was given to both groups [57]. Despite delaying the evolution of infarcts and oedema and decreasing mortality in the early stages, oedema evolved on later days and in the surviving patients, functional outcome was not improved [57].

A new formulation and dosing regimen of edaravone was well tolerated and achieved the intended plasma concentrations [58]. The protocol includes twice daily intravenous infusion over a maximum of 14 days which as this study points out may not be compatible with the much shorter acute hospital stays in Europe [58].

In the latest study, a retrospective cohort using the Japanese Diagnosis Procedure Combination database, the edaravone group ($n=5979$; 94 %) had improved early outcomes as measured by the mRS when compared with the control group ($n=357$; 6 %) [59].

Some studies contained patients who received intravenous thrombolysis in addition to edaravone—depending on clinical cases [47, 56, 57]. Other studies had intravenous thrombolysis as exclusion criteria [45, 54, 55] as reviewed by [44]. However, evidence to suggest a synergistic relationship between edaravone and rtPA is scant (Amaro and Chamorro 2011).

In a multicenter, single-blind, randomised, open-labelled study, patients with M1 or M2 occlusion within 3 h of symptom onset were assigned to 30 mg edaravone (then 30 mg BD for 7 days) and t-PA (23) and t-PA alone (17) [60]. Using the NIHSS score after 8 patients who received endovascular therapy were excluded remarkable and good recoveries were more common in the edaravone group than control (80.1 % vs. 45.5 %, $p=0.0396$) [60]. In a study to look at age and the combination of tPA with and without edaravone 129 patients were divided into two groups according to age 81 patients under 80 years old (76 of these received edaravone) and 48 patients over the age of 80 years old (42 of these received edaravone), age range 34–101 [61]. Poor outcomes were attributes to age and not tPA. It was suggested that a combination of tPA and edaravone augmented recanalization rates and lowered rates of haemorrhagic transformation [61].

There are also trials that look at edaravone in combination with other treatments. The Japanese 'EDO' trial was a multicenter randomised parallel-group open-label trial that enrolled 401 patients and it consisted of intravenous edaravone versus the thromboxane (2) synthase inhibitor, sodium ozagrel in acute non-cardioembolic ischemic stroke [62]. The main conclusion was that edaravone was not inferior to ozagrel, with a downward trend in NIHSS scores ([62] as reviewed in [44]). This was also deemed to be cost effective [63].

In a small randomised Japanese study, in patients with acute embolic stroke in the anterior cerebral circulation within 48 h of symptom onset NIHSS>5, hyperbaric oxygen was combined with intravenous edaravone ($n=19$) and compared with usual treatment control ($n=19$) [64]. The primary endpoint was a favourable mRS at 90 days was significant in the edaravone group. The secondary endpoint NIHSS at 7 days was not, however, significant. The study was very small and effect could not necessarily be attributed to the combination of the two therapies as there was not an edaravone alone and hyperbaric oxygen alone group.

Edaravone is currently only approved in the clinical setting in Japan for use in ischemic stroke within 24 h ([44], Amaro and Chamorro 2011). The mixed results from clinical trials coupled with the heterogeneity of the pathology that they cover explain the insufficiency of the evidence to justify more wide-spread use.

NXY-059

NXY-059 is a nitrone with free radical scavenging ability that fufilled nearly all of the STAIR criteria, yet still failed to make it to clinic. Two clinical trials were undertaken, SAINT I [65] and SAINT II [66]. SAINT I was a phase III double-blinded, randomised, placebo-controlled trial. There was no improvement on neurological outcome (measured by the NIHSS), but improvement in disability (measured by the mRS) did prove to be significant. SAINT II, though statistically more powerful, failed to corroborate the findings of the first study [66]. Reasons why a compound that showed great promise in preclinical studies yet failed to deliver in the clinic include poor penetration of the blood brain barrier, being a weak antioxidant, absence of synergy with rtPA and wide treatment windows (Amaro and Chamorro 2011). In the pooled intention to treat analysis of SAINT I and SAINT II trials (5928 patients; NXY-059 $n=2438$ vs placebo $n=2456$ within 6 h of onset) there was no evidence of improvement of disability (mRS) and mortality was equal in both groups [67].

Antioxidant Nutrients

Low levels of antioxidants have been measured in patients in the acute phase following stroke [68]. There are increased oxidative stress markers present in the serum, months after the initial insult (reviewed by [14]).

Therefore, consideration of long-term damage to antioxidant mechanisms themselves may provide relevant and more precise molecular targets for treatment.

Vitamin C

The AVASAS (Aspirin Versus Ascorbic acid plus Aspirin in Stroke) Study randomised 59 patients admitted within 24 h of symptom onset with ischemic stroke, in an open-label fashion, to orally receive either aspirin (acetylsalicylic acid, ASA) 300 mg plus vitamin C (ascorbic acid) 200 mg/day (28 patients) or ASA 300 mg/day alone (31 patients) for 90 days. Specific plasma markers of lipid peroxidation were measured. Clinically, both groups progressed in a similar manner; by the end of the first week, patients treated with vitamin C plus aspirin had higher vitamin C levels and lower 8,12-isoprostanes F2α-VI (8,12-iPF2α-VI). Levels of vitamin C remained significantly elevated until study completion at 3 months [69].

In a small Polish study, intravenous infusion of 500 mg per day to ischemic stroke patients resulted in elevated serum levels of antioxidants, but it did not substantially improve the clinical and functional status of patients after 3 months [70].

In a randomised control trial containing 96 ischemic stroke patients within 12 h of symptoms starting were assigned to the following groups: daily oral 800 IU (727 mg) vitamin E and 500 mg vitamin C ($n=24$), or B-group vitamins (5 mg folic acid, 5 mg vitamin B2, 50 mg vitamin B6, and 0.4 mg of vitamin B12; $n=24$), both vitamins together ($n=24$), or no supplementation ($n=24$) for 14 days [71]. Based on plasma blood levels without imaging or functional outcomes, antioxidant capacity was augmented, oxidative damage diminished and there were potential anti-inflammatory effects.

In a retrospective case–control study of 23 patients taking vitamin C (prescribed to those who were undernourished) following ischemic stroke and 23 patients who were not, functional recovery was not improved by vitamin C supplementation [72].

Vitamin E

Vitamin E has been mooted as a therapeutic agent in other neurological disorders. In patients with mild to moderate Alzheimer's disease vitamin E appears to slow functional decline [73].

A systematic review and meta-analysis of randomised, placebo-controlled trials using vitamin E published until January 2010. There was a relatively small risk reduction for ischemic stroke and a large increase in the risk of haemorrhagic stroke [74].

These equivocal results in stroke are in keeping with other clinical trials concerning cardiovascular disease. In a trial of over, 9000 patients, there was no benefit of vitamin E supplementation in high-risk subjects over a 4.5 year study period [75]. In addition, vitamin E has not been shown to be beneficial for the treatment of mild cognitive impairment [76].

Other Compounds

The compounds discussed thus far target oxidative stress very directly, either through scavenging or inhibition of In addition, other compounds exist that may exert their effects through antioxidant mechanisms albeit via a more indirect mechanism of action.

Statins

Statins are 3-hydroxy-3-methyl-glutaryl-CoA reductase (HMG-CoA) reductase inhibitors, aside from lowering cholesterol they have other pleiotropic effects one of which is the ability to act as an antioxidant. Statins inhibit NADPH oxidase-mediated production of free radicals by multiple mechanisms. First, they reduce the isoprenylation of its activator, the GTP binding protein rac1, by inhibiting its translocation to the membrane. Second, they decrease the expression of angiotensin-1 receptor (AT-1) at which angiotensin II binds and is a positive controller of rac1 [77, 78]. Another mechanism is via control of NO by encouraging normal production of NO via eNOS and by preventing ischemia-induced activation of nNOS and iNOS (as reviewed by [4]). eNOS-derived NO exhibits a variety of beneficial effects including enhanced collateral flow due to vasodilation, reduced inflammation and neovascularisation in a more chronic context; in contrast, iNOS/nNOS-derived NO can lead to peroxynitrite production [80]

In a small study, of 67 patients, an antioxidant effect of statins was inferred by measuring 8-isoprostane levels with the median levels being significantly lower in those receiving simvastatin [81]. However, it is difficult to know how much of the effect of a particular drug can be attributed to a specific mechanism of action. In a recent Cochrane Review it was concluded that there was insufficient data from randomised trials to establish the safety and efficacy of statins in acute ischemic stroke and TIA [82].

Tirilazad

The non-glucocorticoid 21-aminosteroid lipid peroxidation inhibitor, tirilazad, although demonstrating promise in preclinical trails, did not improve functional outcome in clinical trials [83, 84].

NA-1

Lessons should also be learned from NA-1 (Tat-NR2B9c). NA-1 is a post-synaptic density-95 (PSD-95) inhibitor and not a direct inhibitor of oxidative stress, although it does prevent NO/RNS formation as measured by cGMP production [85, 86]. This is achieved by uncoupling nNOS from glutamate receptor activation, thus reducing harmful NO production in the context of excitotoxicity.

The phase 2, randomised, double-blind, placebo-controlled trial included patients aged 18 or over with ruptured or unruptured intracranial aneurysms with the potential to repair. There were 185 eligible patients NA-1 $n=92$ and placebo $n=93$. There was no significant difference in lesion volume assessed by diffusion-weighted MRI or fluid-attenuated inversion recovery MRI (adjusted p value $=0 \cdot 236$). A claim was made about the neuroprotective potential for NA-1 based on the experimental group having fewer ischemic infarcts when compared to placebo [87]; however, there has yet to be a definitive trial on the effects of NA-1 in acute ischaemic stroke.

The Future

NADPH Oxidases

Although there are currently no compounds undergoing clinical trials, exploiting a pathway dedicated to ROS production has yielded small molecule sub-type specific NADPH oxidase inhibitors that are under development [16, 88].

Neurovascular Unit

Increasingly, the emphasis in stroke research is on the neurovascular unit (NVU), a conceptual approach that addresses interactions between the vasculature, glia and neurons instead of targeting a single cell type [89]. This is particularly relevant for oxidative stress research, as stroke causes superoxide- and peroxynitrite-mediated vascular damage and blood–brain barrier breakdown [90]. Indeed, the promising candidate neuroprotectant edaravone exhibits protective effects not only on neurons, but also reduces oxidative stress damage in astrocyte and endothelial cultures [91]. It is not currently possible to distinguish between direct neuronal and broader neurovascular effects in the context of clinical trials, but the preclinical evidence suggests that the putative beneficial effects of antioxidants could be at least partly due to protection of the NVU.

One mechanism by which antioxidants might exert protective effects is through restoration of microvascular patency. Recanalisation does not always result in reperfusion of the affected tissue, which diminishes the potential benefit of thrombolysis [92]. Similarly, no-reflow and post-ischemic hypoperfusion are a well-established concept in preclinical studies, where opening the experimentally occluded arteries does not fully restore flow to the developing infarct [93, 94]. There are multiple explanations for these phenomena, including adhesion of leukocytes and platelets in the affected region leading to obstruction of capillaries and venules, altered vascular tone and swelling of astrocytic endfeet surrounding the microvessels [95]. More recently, the constriction of pericytes—vascular mural cells closely surrounding the endothelium—has been implicated in no-reflow. Notably, preventing oxidative and nitrative stress through scavenger compounds restores capillary patency in mice following stroke [96]. In cerebellar slices, ischemia resulted in pericyte constriction and death in rigour, which could be reduced by nitric oxide synthase blockers but not by ROS scavengers [97]. Nonetheless, these recent publications highlight the importance of neurovascular function following stroke, and indicate potential roles for antioxidants in achieving reperfusion.

However, the picture is further complicated by reperfusion injury. Although the clinical benefit of reperfusion is clearly supported by a great deal of evidence, we also know that the NVU is susceptible to further damage when flow is restored, and in some animal models of late reperfusion there is a marked increase in blood–brain barrier damage, oedema and infarct size [98, 99]. This is again a multifactorial complication, featuring multiple forms of cell death and inflammatory damage. Inflammation in particular is potentially amenable to antioxidant therapies, given the propensity of inflammatory cells to produce various free radical species [100]. The exact radical species involved are uncertain, with both oxygen and nitrogen radicals being implicated; in contrast, some studies have reported beneficial effects from NO donors through ROS scavenging following ischemia/reperfusion [101].

Physiological Parameters

In addition to molecular targets, optimisation of physiological parameters should also be considered. A recent trial in 36 patients 1-h post-thrombolysis with a median NIHSS of 9 has demonstrated that poor outcome (mRS, 4–6) was twice as common in the normothermia group 44 % versus 22 % in the mild hypothermia (35 °C) group [102]. The management of blood glucose is paramount in the management of acute ischemic stroke as hyperglycaemia is associated with greater cortical toxicity and larger infarct volumes [103] and adversely affect outcome [104].

Conclusion

As yet, there have been no unequivocally successful clinical applications of antioxidants to scavenge reactive oxygen or nitrogen species (ROS/RNS), with some even having a detrimental effect. Although ROS and RNS are molecules that are only present transiently, they still have the ability to cause a huge amount of damage. The key to targeting this mechanism of cerebral damage should therefore lie in preventing their formation in the first place by targeting the enzymes responsible for their production e.g. NADPH oxidases [105].

One of the major scientific goals in the neurosciences, has been to develop so-called neuroprotectants which, if they worked in man, would increase the amount of time we had to image, treat, and restore blood flow to the brain and, if possible, would protect the brain against ischemia, resulting in less damage. Studies have so far concentrated on suppressing the "ischemic cascade" which at best amounts to 'damage limitation'. Discovering and amplifying endogenous protective mechanisms mediated by proteins such as hamartin could produce a tangible clinical effect by preventing ischemic damage in the first place [106].

Stroke is an insult to the neurovascular unit (i.e. neurons in combination with endothelial cells, astrocytes and pericytes) not isolated neurons therapies, and targeted as such may have more success in future trials (reviewed by [14]). Fifteen years ago, the Stroke Therapy Industry Roundtable (STAIR) emphasised the importance of targeting multiple mechanisms and multiple cell types in the treatment of stroke [107].

When looking for compounds in preclinical studies to carry forward to clinical trials it is paramount to distinguish between true neuroprotective effects at the proposed target and secondary alterations of physiological parameters. For example MK-801 conferred neuroprotection via hypothermia [109] and potentially an alteration of cerebral blood flow [109]—not through its actions at the NMDA receptor. It is important to use novel imaging techniques in order to demonstrate causality, such as visualising oxidative stress and subsequent mitochondrial dysfunction using two-photon microscopy (as reviewed by Sutherland et al) [16, 110]. Changes in glutathione levels between astrocytes and neurons have also been visualised in ischemic stroke [111].

Perhaps it is time to approach preclinical studies like clinical trials with multi-center randomised double-blind placebo-controlled trials in realistic animal models of stroke in order to ensure a greater correlation between the success of preclinical findings that has so far eluded clinical studies [112, 113]. There is also a need for concordance when it comes to endpoints, so that they are meaningful.

Targeting oxidative stress is unlikely to provide the panacea for ischemic stroke. Indeed with some of the proposed therapies it is hard to know the extent to which they affect oxidative stress. Future successful treatments are likely to be combination therapies that not only complement each other, but have synergistic mechanisms of action (Amaro and Chamorro 2011) and targeting the ischemic cascade and resultant oxidative stress at its origin is likely to be more effective than mere damage limitation (Table 7.1).

Table 7.1 Randomised double-blind clinical trials looking at antioxidant therapy in ischemic stroke

Drug	Trial	Mechanism	Clinical phase	Manufacturer	Summary
Completed					
Tirilazad mesylate	[83]	Lipid peroxidation inhibitor	Safety and efficacy	'The Upjohn Company'	660 patients were randomised. 556 eligible (276 tirilazad, 280 vehicle). Treatment within 6 h of onset tirilazad, 6 mg/kg per day for 3 days. Terminated early as it did not improve overall functional outcome
NXY-059 (Cerovive)	[65]	Free radical scavenger		AstraZeneca	$n=1722$ NXY-059 did produce significant benefits ($p=0.038$) in the modified Rankin functional scale but not NIHSS or Barthel Index.
NXY-059 (Cerovive)	[66]			AstraZeneca	$n=3206$ failed to reach significance.
Edaravone (Radicut®) (MCI-186)	[46]	Free radical scavenger	Safety and efficacy	Mitsubishi Pharma Corporation	50 patients. Edaravone 18/25 (72 %) had favourable outcomes (MRS ≤ 2) at 90 days vs 10/25 (40 %) in placebo group ($p<0.005$). Two patients died (one in each group) during treatment. Day 90 Mean Barthel index: Edaravone increased from 41.20 ± 32.70 at baseline to 82.40 ± 18.32 at day 90 vs placebo group 44.20 ± 22.76 at baseline and 68.20 ± 21.30 at day 90 ($p<0.005$)
Edaravone (Radicut®) (MCI-186)		Free radical scavenger	IIa	Mitsubishi Pharma Corporation	Multi-centre, randomised, double-blind, placebo-controlled, clinical study which has been completed but not yet reported, looking at patients with acute ischemic stroke within 24 h of onset using either 1000 or 2000 mg edaravone http://www.clinicaltrials.gov/ct2/show/NCT00821821
Ongoing					
Ebselen		Free radical scavenger	III	Daiichi Pharmaceutical Co., LTD	
Edaravone (Radicut®) (MCI-186)		Free radical scavenger	III	Mitsubishi Pharma Corporation	

Edaravone (Radicut®) (MCI-186)		Free radical scavenger	Retrospective comparison with a historical-controlled cohort	Mitsubishi Pharma Corporation	Early improvement (between day 0 and day 10): NIHSS on admission < or =7) among the ED vs control group (change in NIHSS +2 vs. −2, respectively, $p=0.013$). Moderate to severe (NIHSS > 7) patients, not significant
Uric acid (in 500 ml 0.1 % lithium carbonate and 5 % Mannitol), given with rtPA		Free radical scavenger	III	Uric acid obtained from Sigma Trial: efficacy Study of Combined Treatment With Uric Acid and rtPA in Acute Ischemic Stroke (URICO-ICTUS)	Estimated enrolment: 420 patients over 3 years, from Jan 2010. 1 g of uric acid dissolved in vehicle (500 ml of 0.1 % lithium carbonate and 5 % mannitol) ($n=210$) or vehicle alone ($n=210$). Patients ≤18 years, treated with rtPA within the first 4.5 h of onset. Baseline NIHSS > 6 and <25 and a mRS ≤ 2 Primary outcome: proportion patients mRS 0–1 at 3 months or 2 in those with a prior qualifying mRS of 2

References

1. Geeganage C, Bath PM. Interventions for deliberately altering blood pressure in acute stroke. Cochrane Database Syst Rev. 2008;4:CD000039.
2. The National Institute of Neurological Disorders and Stroke rt-PA Stroke Study Group. Tissue plasminogen activator for acute ischemic stroke. N Engl J Med. 1995;333(24): 1581–7.
3. Balami JS, Sutherland BA, Edmunds LD, Grunwald IQ, Neuhaus AA, Hadley G, et al. A systematic review and meta-analysis of randomized controlled trials of endovascular thrombectomy compared with best medical treatment for acute ischemic stroke. Int J Stroke. 2015;10(8):1168–78.
4. Cimino M, Gelosa P, Gianella A, Nobili E, Tremoli E, Sironi L. Statins: multiple mechanisms of action in the ischemic brain. Neuroscientist. 2007;13(3):208–13.
5. Crack PJ, Taylor JM. Reactive oxygen species and the modulation of stroke. Free Radic Biol Med. 2005;38(11):1433–44. Review.
6. Flamm ES, Demopoulos HB, Seligman ML, Poser RG, Ransohoff J. Free radicals in cerebral ischemia. Stroke. 1978;9(5):445–7.
7. Miller AA, Drummond GR, Sobey CG. Novel isoforms of NADPH-oxidase in cerebral vascular control. Pharmacol Ther. 2006;111(3):928–48. Epub 2006 Apr 17.
8. Orrenius S, Zhivotovsky B, Nicotera P. Regulation of cell death: the calcium-apoptosis link. Nat Rev Mol Cell Biol. 2003;4(7):552–65.
9. McCord JM. Oxygen-derived free radicals in postischemic tissue injury. N Engl J Med. 1985;312(3):159–63.
10. Neuhaus A et al. Importance of preclinical research in the development of neuroprotective strategies for ischemic stroke. JAMA Neurol. 2014;71:634–9.
11. Li C, Jackson RM. Reactive species mechanisms of cellular hypoxia-reoxygenation injury. Am J Physiol Cell Physiol. 2002;282(2):C227–41.
12. Hossmann K-A. The two pathophysiologies of focal brain ischemia: implications for translational stroke research. J Cereb Blood Flow Metab. 2012;32(7):1310–6.
13. Shichinohe H et al. Neuroprotective effects of the free radical scavenger edaravone (MCI-186) in mice permanent focal brain ischemia. Brain Res. 2004;1029(2):200–6.
14. Manzanero S, Santro T, Arumugam TV. Neuronal oxidative stress in acute ischemic stroke: sources and contribution to cell injury. Neurochem Int. 2013;62(5):712–8. doi:10.1016/j.neuint.2012.11.009. Epub 2012 Nov 29. Review.
15. Pál P, Nivorozhkin A, Szabó C. Therapeutic effects of xanthine oxidase inhibitors: renaissance half a century after the discovery of allopurinol. Pharmacol Rev. 2006;58(1):87–114.
16. Sutherland BA, Minnerup J, Balami JS, Arba F, Buchan AM, Kleinschnitz C. Neuroprotection for ischaemic stroke: translation from the bench to the bedside. Int J Stroke. 2012;7(5):407–18. doi:10.1111/j.1747-4949.2012.00770.x. Epub 2012 Mar 6.
17. Ames BN, Cathcart R, Schwiers E, Hochstein P. Uric acid provides an antioxidant defense in humans against oxidant- and radical-caused aging and cancer: a hypothesis. Proc Natl Acad Sci U S A. 1981;78(11):6858–62.
18. Amaro S, Chamorro A. Translational stroke research of the combination of thrombolysis and antioxidant therapy. Stroke. 2011;42(5):1495–9.
19. Du Y, Chen CP, Tseng CY, Eisenberg Y, Firestein BL. Astroglia-mediated effects of uric acid to protect spinal cord neurons from glutamate toxicity. Glia. 2007;55(5):463–72.
20. Holme I, Aastveit AH, Hammar N, Jungner I, Walldius G. Uric acid and risk of myocardial infarction, stroke and congestive heart failure in 417,734 men and women in the Apolipoprotein MOrtality RISk study (AMORIS). J Intern Med. 2009;266(6):558–70. doi:10.1111/j.1365-2796.2009.02133.x. Epub 2009 May 26.
21. Kim SY, Guevara JP, Kim KM, Choi HK, Heitjan DF, Albert DA. Hyperuricemia and risk of stroke: a systematic review and meta-analysis. Arthritis Rheum. 2009;61(7):885–92. doi:10.1002/art.24612.

22. Chen JH, Chuang SY, Chen HJ, Yeh WT, Pan WH. Serum uric acid level as an independent risk factor for all-cause, cardiovascular, and ischemic stroke mortality: a Chinese cohort study. Arthritis Rheum. 2009;61(2):225–32. doi:10.1002/art.24164.
23. Chamorro A, Planas AM, Muner DS, Deulofeu R. Uric acid administration for neuroprotection in patients with acute brain ischemia. Med Hypotheses. 2004;62(2):173–6.
24. Culleton BF, Larson MG, Kannel WB, Levy D. Serum uric acid and risk for cardiovascular disease and death: the Framingham Heart Study. Ann Intern Med. 1999;131(1):7–13.
25. Waring WS. Uric acid: an important antioxidant in acute ischaemic stroke. QJM. 2002; 95(10):691–3.
26. Bos MJ, Koudstaal PJ, Hofman A, Witteman JC, Breteler MM. Uric acid is a risk factor for myocardial infarction and stroke: the Rotterdam study. Stroke. 2006;37(6):1503–7. Epub 2006 May 4.
27. Storhaug HM, Norvik JV, Toft I, Eriksen BO, Løchen ML, Zykova S, et al. Uric acid is a risk factor for ischemic stroke and all-cause mortality in the general population: a gender specific analysis from The Tromsø Study. BMC Cardiovasc Disord. 2013;13:115. doi:10.1186/1471-2261-13-115.
28. Weir CJ, Muir SW, Walters MR, Lees KR. Serum urate as an independent predictor of poor outcome and future vascular events after acute stroke. Stroke. 2003;34(8):1951–6. Epub 2003 Jul 3.
29. Dawson J, Lees KR, Weir CJ, Quinn T, Ali M, Hennerici MG, et al. Baseline serum urate and 90-day functional outcomes following acute ischemic stroke. Cerebrovasc Dis. 2009;28(2):202–3. doi:10.1159/000226580. Epub 2009 Jun 30.
30. Dawson J, Quinn TJ, Harrow C, Lees KR, Walters MR. The effect of allopurinol on the cerebral vasculature of patients with subcortical stroke; a randomized trial. Br J Clin Pharmacol. 2009;68(5):662–8. doi:10.1111/j.1365-2125.2009.03497.x.
31. Kurzepa J, Bielewicz J, Stelmasiak Z, Bartosik-Psujek H. Serum bilirubin and uric acid levels as the bad prognostic factors in the ischemic stroke. Int J Neurosci. 2009;119(12):2243–9.
32. Chiquete E, Ruiz-Sandoval JL, Murillo-Bonilla LM, Arauz A, Orozco-Valera DR, Ochoa-Guzmán A, et al. Serum uric acid and outcome after acute ischemic stroke: PREMIER study. Cerebrovasc Dis. 2013;35(2):168–74. doi:10.1159/000346603. Epub 2013 Feb 22.
33. Chamorro A, Obach V, Cervera A, Revilla M, Deulofeu R, Aponte JH. Prognostic significance of uric acid serum concentration in patients with acute ischemic stroke. Stroke. 2002;33(4):1048–52.
34. Zhang B, Gao C, Yang N, Zhang W, Song X, Yin J, et al. Is elevated SUA associated with a worse outcome in young Chinese patients with acute cerebral ischemic stroke? BMC Neurol. 2010;10:82. doi:10.1186/1471-2377-10-82.
35. Lee SH, Heo SH, Kim JH, Lee D, Lee JS, Kim YS, et al. Effects of uric acid levels on outcome in severe ischemic stroke patients treated with intravenous recombinant tissue plasminogen activator. Eur Neurol. 2013;71(3-4):132–9 [Epub ahead of print].
36. Chamorro Á, Amaro S, Castellanos M, Segura T, Arenillas J, Martí-Fábregas J, et al. Safety and efficacy of uric acid in patients with acute stroke (URICO-ICTUS): a randomised, double-blind phase 2b/3 trial. Lancet Neurol. 2014;13(5):453–60.
37. Romanos E, Planas AM, Amaro S, Chamorro A. Uric acid reduces brain damage and improves the benefits of rt-PA in a rat model of thromboembolic stroke. J Cereb Blood Flow Metab. 2007;27(1):14–20. Epub 2006 Apr 5.
38. Amaro S, Obach V, Cervera A, Urra X, Gómez-Choco M, Planas AM, et al. Course of matrix metalloproteinase-9 isoforms after the administration of uric acid in patients with acute stroke: a proof-of-concept study. J Neurol. 2009;256(4):651–6. doi:10.1007/s00415-009-0153-6. Epub 2009 Apr 27.
39. Amaro S, Soy D, Obach V, Cervera A, Planas AM, Chamorro A. A pilot study of dual treatment with recombinant tissue plasminogen activator and uric acid in acute ischemic stroke. Stroke. 2007;38(7):2173–5. Epub 2007 May 24.
40. Taheraghdam AA, Sharifipour E, Pashapour A, Namdar S, Hatami A, Houshmandzad S, et al. Allopurinol as a preventive contrivance after acute ischemic stroke in patients with a high

level of serum uric acid: a randomized, controlled trial. Med Princ Pract. 2014;23:134–9. doi:10.1159/000355621. Epub 2013 Nov 27.
41. Khan F, George J, Wong K, McSwiggan S, Struthers AD, Belch JJ. Allopurinol treatment reduces arterial wave reflection in stroke survivors. Cardiovasc Ther. 2008;26(4):247–52. doi:10.1111/j.1755-5922.2008.00057.x.
42. Muir SW, Harrow C, Dawson J, Lees KR, Weir CJ, Sattar N, et al. Allopurinol use yields potentially beneficial effects on inflammatory indices in those with recent ischemic stroke: a randomized, double-blind, placebo-controlled trial. Stroke. 2008;39(12):3303–7. doi:10.1161/STROKEAHA.108.519793. Epub 2008 Oct 9.
43. Proctor PH. Uric acid and neuroprotection. Stroke. 2008;39(8), e126. doi:10.1161/STROKEAHA.108.524462. Epub 2008 Jun 19.
44. Lapchak PA. A critical assessment of edaravone acute ischemic stroke efficacy trials: is edaravone an effective neuroprotective therapy? Expert Opin Pharmacother. 2010; 11(10):1753–63. doi:10.1517/14656566.2010.493558.
45. Edaravone Acute Infarction Study Group. Effect of a novel free radical scavenger, edaravone (MCI-186), on acute brain infarction. Randomized, placebo-controlled, double-blind study at multicenters. Cerebrovasc Dis. 2003;15(3):222–9.
46. Sharma P, Sinha M, Shukla R, Garg RK, Verma R, Singh MK. A randomized controlled clinical trial to compare the safety and efficacy of edaravone in acute ischemic stroke. Ann Indian Acad Neurol. 2011;14(2):103–6. doi:10.4103/0972-2327.82794.
47. Unno Y, Katayama M, Shimizu H. Does functional outcome in acute ischaemic stroke patients correlate with the amount of free-radical scavenger treatment? A retrospective study of edaravone therapy. Clin Drug Investig. 2010;30(3):143–55. doi:10.2165/11535500-000000000-00000.
48. Zhang M, Xu L, Deng L, Lu J, Ren H, Yang Q, et al. Efficacy and safety evaluation of edaravone injection in treatment of acute cerebral infarction: a multicenter, double-blind, and randomized controlled clinical trial. Chin J New Drugs Clin Remedies. 2007;26(2):105–8.
49. Zhou M, Yang J, He L. Randomized controlled trial of edaravone injection in the treatment of acute cerebral infarction. Modern Preventive Medicine. 2007;34:966–8.
50. Feng S, Yang Q, Liu M, Li W, Yuan W, Zhang S, et al. Edaravone for acute ischaemic stroke. Cochrane Database Syst Rev. 2011;12:007230. doi:10.1002/14651858.CD007230.pub2.
51. Yang J, Cui X, Li J, Zhang C, Zhang J, Liu M. Edaravone for acute stroke: meta-analyses of data from randomized controlled trials. Dev Neurorehabil. 2013;2 [Epub ahead of print].
52. Ishibashi A, Yoshitake Y, Adachi H. Investigation of effect of edaravone on ischemic stroke. Kurume Med J. 2013;2 [Epub ahead of print].
53. Siesjo BK, Agardh CD, Bengtsson F. Free radicals and brain damage. Cerebrovasc Brain Metab Rev. 1989;1:165–211.
54. Mishina M, Komaba Y, Kobayashi S, Tanaka N, Kominami S, Fukuchi T, et al. Efficacy of edaravone, a free radical scavenger, for the treatment of acute lacunar infarction. Neurol Med Chir (Tokyo). 2005;45(7):344–8. discussion 348.
55. Ohta Y, Takamatsu K, Fukushima T, Ikegami S, Takeda I, Ota T, et al. Efficacy of the free radical scavenger, edaravone, for motor palsy of acute lacunar infarction. Intern Med. 2009;48(8):593–6. Epub 2009 Apr 15.
56. Inatomi Y, Takita T, Yonehara T, Fujioka S, Hashimoto Y, Hirano T, et al. Efficacy of edaravone in cardioembolic stroke. Intern Med. 2006;45(5):253–7. Epub 2006 Apr 3.
57. Toyoda K, Fujii K, Kamouchi M, Nakane H, Arihiro S, Okada Y, et al. Free radical scavenger, edaravone, in stroke with internal carotid artery occlusion. J Neurol Sci. 2004;221(1-2): 11–7.
58. Kaste M, Murayama S, Ford GA, Dippel DW, Walters MR, Tatlisumak T, et al. Safety, tolerability and pharmacokinetics of MCI-186 in patients with acute ischemic stroke: new formulation and dosing regimen. Cerebrovasc Dis. 2013;36(3):196–204. doi:10.1159/000353680. Epub 2013 Oct 12.
59. Wada T, Yasunaga H, Inokuchi R, Horiguchi H, Fushimi K, Matsubara T, et al. Effects of edaravone on early outcomes in acute ischemic stroke patients treated with recombinant tissue plasminogen activator. J Neurol Sci. 2014;pii:S0022-510X(14)00464-X. doi:10.1016/j.jns.2014.07.018.

60. Kimura K, Aoki J, Sakamoto Y, Kobayashi K, Sakai K, Inoue T, et al. Administration of edaravone, a free radical scavenger, during t-PA infusion can enhance early recanalization in acute stroke patients—a preliminary study. J Neurol Sci. 2012;313(1-2):132–6. doi:10.1016/j.jns.2011.09.006. Epub 2011 Oct 2.
61. Kono S, Deguchi K, Morimoto N, Kurata T, Yamashita T, Ikeda Y, et al. Intravenous thrombolysis with neuroprotective therapy by edaravone for ischemic stroke patients older than 80 years of age. J Stroke Cerebrovasc Dis. 2013;22(7):1175–83.
62. Shinohara Y, Saito I, Kobayashi S, Uchiyama S. Edaravone (radical scavenger) versus sodium ozagrel (antiplatelet agent) in acute noncardioembolic ischemic stroke (EDO trial). Cerebrovasc Dis. 2009;27(5):485–92. doi:10.1159/000210190. Epub 2009 Mar 26.
63. Shinohara Y, Inoue S. Cost-effectiveness analysis of the neuroprotective agent edaravone for noncardioembolic cerebral infarction. J Stroke Cerebrovasc Dis. 2013;22(5):668–74. doi:10.1016/j.jstrokecerebrovasdis.2012.04.002.
64. Imai K, Mori T, Izumoto H, Takabatake N, Kunieda T, Watanabe M. Hyperbaric oxygen combined with intravenous edaravone for treatment of acute embolic stroke: a pilot clinical trial. Neurol Med Chir (Tokyo). 2006;46(8):373–8. discussion 378.
65. Lees KR, Zivin JA, Ashwood T, Davalos A, Davis SM, Diener HC, et al. NXY-059 for acute ischemic stroke. N Engl J Med. 2006;354(6):588–600.
66. Shuaib A, Lees KR, Lyden P, Grotta J, Davalos A, Davis SM, et al. NXY-059 for the treatment of acute ischemic stroke. N Engl J Med. 2007;357(6):562–71.
67. Diener HC, Lees KR, Lyden P, Grotta J, Davalos A, Davis SM, et al. NXY-059 for the treatment of acute stroke: pooled analysis of the SAINT I and II Trials. Stroke. 2008;39(6):1751–8. doi:10.1161/STROKEAHA.107.503334.
68. Sharpe PC, Mulholland C, Trinick T. Ascorbate and malondialdehyde in stroke patients. Ir J Med Sci. 1994;163(11):488–91.
69. Polidori MC, Praticó D, Ingegni T, Mariani E, Spazzafumo L, Del Sindaco P, et al. Effects of vitamin C and aspirin in ischemic stroke-related lipid peroxidation: results of the AVASAS (Aspirin Versus Ascorbic acid plus Aspirin in Stroke) Study. Biofactors. 2005;24(1-4): 265–74.
70. Lagowska-Lenard M, Stelmasiak Z, Bartosik-Psujek H. Influence of vitamin C on markers of oxidative stress in the earliest period of ischemic stroke. Pharmacol Rep. 2010;62(4):751–6.
71. Ullegaddi R, Powers HJ, Gariballa SE. Antioxidant supplementation with or without B-group vitamins after acute ischemic stroke: a randomized controlled trial. JPEN J Parenter Enteral Nutr. 2006;30(2):108–14.
72. Rabadi MH, Kristal BS. Effect of vitamin C supplementation on stroke recovery: a case-control study. Clin Interv Aging. 2007;2(1):147–51.
73. Dysken MW, Sano M, Asthana S, Vertrees JE, Pallaki M, Llorente M, et al. Effect of vitamin E and memantine on functional decline in Alzheimer disease: the TEAM-AD VA cooperative randomized trial. JAMA. 2014;311(1):33–44. doi:10.1001/jama.2013.282834.
74. Schürks M, Glynn RJ, Rist PM, Tzourio C, Kurth T. Effects of vitamin E on stroke subtypes: meta-analysis of randomised controlled trials. BMJ. 2010;341:c5702. doi:10.1136/bmj.c5702.
75. Yusuf S, Dagenais G, Pogue J, Bosch J, Sleight p. Vitamin E supplementation and cardiovascular events in high-risk patients. The Heart Outcomes Prevention Evaluation Study Investigators. N Engl J Med. 2000;342(3):154–60.
76. Petersen RC, Thomas RG, Grundman M, Bennett D, Doody R, Ferris S, et al. Vitamin E and donepezil for the treatment of mild cognitive impairment. N Engl J Med. 2005; 352(23):2379–88.
77. Wagner AH, Köhler T, Rückschloss U, Just I, Hecker M. Improvement of nitric oxide-dependent vasodilatation by HMG-CoA reductase inhibitors through attenuation of endothelial superoxide anion formation. Arterioscler Thromb Vasc Biol. 2000;20(1):61–9.
78. Wassmann S, Laufs U, Bäumer AT, Müller K, Konkol C, Sauer H, et al. Inhibition of geranylgeranylation reduces angiotensin II-mediated free radical production in vascular smooth muscle cells: involvement of angiotensin AT1 receptor expression and Rac1 GTPase. Mol Pharmacol. 2001;59(3):646–54.

80. Endres M, Laufs U, Liao JK, Moskowitz MA. Targeting eNOS for stroke protection. Trends Neurosci. 2004;27(5):283–9.
81. Szczepańska-Szerej A, Kurzepa J, Wojczal J, Stelmasiak Z. Simvastatin displays an antioxidative effect by inhibiting an increase in the serum 8-isoprostane level in patients with acute ischemic stroke: brief report. Clin Neuropharmacol. 2011;34(5):191–4. doi:10.1097/WNF.0b013e3182309418.
82. Squizzato A, Romualdi E, Dentali F, Ageno W. Statins for acute ischemic stroke. Cochrane Database Syst Rev. 2011;8, CD007551. doi:10.1002/14651858.CD007551.pub2.
83. The RANTTAS Investigators. A randomized trial of tirilazad mesylate in patients with acute stroke (RANTTAS). Stroke. 1996;27(9):1453–8. http://www.ncbi.nlm.nih.gov/pubmed/8784112.
84. Sena E, Wheble P, Sandercock P, Macleod M. Systematic review and meta-analysis of the efficacy of tirilazad in experimental stroke. Stroke. 2007;38(2):388–94. Epub 2007 Jan 4.
85. Aarts M, Liu Y, Liu L, Besshoh S, Arundine M, Gurd JW, et al. Treatment of ischemic brain damage by perturbing NMDA receptor- PSD-95 protein interactions. Science. 2002; 298(5594):846–50.
86. Sattler R, Xiong Z, Lu WY, Hafner M, MacDonald JF, Tymianski M. Specific coupling of NMDA receptor activation to nitric oxide neurotoxicity by PSD-95 protein. Science. 1999;284(5421):1845–8.
87. Hill MD, Martin RH, Mikulis D, Wong JH, Silver FL, Terbrugge KG, et al. Safety and efficacy of NA-1 in patients with iatrogenic stroke after endovascular aneurysm repair (ENACT): a phase 2, randomised, double-blind, placebo-controlled trial. Lancet Neurol. 2012;11(11): 942–50. doi:10.1016/S1474-4422(12)70225-9.
88. Kahles T, Brandes RP. NADPH oxidases as therapeutic targets in ischemic stroke. Cell Mol Life Sci. 2012;69(14):2345–63. doi:10.1007/s00018-012-1011-8. Epub 2012 May 23.
89. Arai K et al. Cellular mechanisms of neurovascular damage and repair after stroke. J Child Neurol. 2011;26(9):1193–8.
90. Gürsoy-Ozdemir Y, Can A, Dalkara T. Reperfusion-induced oxidative/nitrative injury to neurovascular unit after focal cerebral ischemia. Stroke. 2004;35(6):1449–53.
91. Lee BJ et al. Edaravone, a free radical scavenger, protects components of the neurovascular unit against oxidative stress in vitro. Brain Res. 2010;1307:22–7.
92. Soares BP et al. Reperfusion is a more accurate predictor of follow-up infarct volume than recanalization: a proof of concept using CT in acute ischemic stroke patients. Stroke. 2010;41(1):e34–40.
93. Ames A et al. Cerebral ischemia. II. The no-reflow phenomenon. Am J Pathol. 1968;52(2): 437–53.
94. Del Zoppo GJ. Toward the neurovascular unit. A journey in clinical translation: 2012 Thomas Willis Lecture. Stroke. 2013;44(1):263–9.
95. Hossmann KA. Reperfusion of the brain after global ischemia: hemodynamic disturbances. Shock (Augusta, GA). 1997;8(2):95–101. discussion 102–3.
96. Yemisci M et al. Pericyte contraction induced by oxidative-nitrative stress impairs capillary reflow despite successful opening of an occluded cerebral artery. Nat Med. 2009;15(9): 1031–7.
97. Hall CN et al. Capillary pericytes regulate cerebral blood flow in health and disease. Nature. 2014;508(7494):55–60.
98. Aronowski J, Strong R, Grotta JC. Reperfusion injury: demonstration of brain damage produced by reperfusion after transient focal ischemia in rats. J Cereb Blood Flow Metab. 1997;17(10):1048–56.
99. Yang GY, Betz L. Reperfusion-induced injury to the blood-brain barrier after middle cerebral artery occlusion in rats. Stroke. 1994;25(8):1658–64.
100. Eltzschig HK, Eckle T. Ischemia and reperfusion—from mechanism to translation. Nat Med. 2011;17(11):1391–401.
101. Pluta RM et al. Effects of nitric oxide on reactive oxygen species production and infarction size after brain reperfusion injury. Neurosurgery. 2001;48(4):884–92. discussion 892–3.

102. Piironen K, Tiainen M, Mustanoja S, Kaukonen KM, Meretoja A, Tatlisumak T, et al. Mild hypothermia after intravenous thrombolysis in patients with acute stroke: a randomized controlled trial. Stroke. 2014;45(2):486–91. doi:10.1161/STROKEAHA.113.003180. Epub 2014 Jan 16.
103. Southerland AM, Johnston KC. Considering hyperglycemia and thrombolysis in the Stroke Hyperglycemia Insulin Network Effort (SHINE) trial. Ann N Y Acad Sci. 2012;1268:72–8. doi:10.1111/j.1749-6632.2012.06731.x.
104. Radermecker RP, Scheen AJ. Management of blood glucose in patients with stroke. Diabetes Metab. 2010;36 Suppl 3:S94–9. doi:10.1016/S1262-3636(10)70474-2.
105. Kleinschnitz C, Grund H, Wingler K, Armitage ME, Jones E, Mittal M, et al. Post-stroke inhibition of induced NADPH oxidase type 4 prevents oxidative stress and neurodegeneration. PLoS Biol. 2010;8(9):pii:1000479. doi:10.1371/journal.pbio.1000479.
106. Papadakis M, Hadley G, Xilouri M, Hoyte LC, Nagel S, McMenamin MM, et al. Tsc1 (hamartin) confers neuroprotection against ischemia by inducing autophagy. Nat Med. 2013;19(3):351–7. doi:10.1038/nm.3097. Epub 2013 Feb 24.
107. Stroke Therapy Academic Industry Roundtable. Recommendations for standards regarding preclinical neuroprotective and restorative drug development. Stroke. 1999;30(12):2752–8. Review.
108. Buchan A, Pulsinelli WA. Hypothermia but not the N-methyl-D-aspartate antagonist, MK-801, attenuates neuronal damage in gerbils subjected to transient global ischemia. J Neurosci. 1990;10(1):311–6.
109. Buchan AM, Slivka A, Xue D. The effect of the NMDA receptor antagonist MK-801 on cerebral blood flow and infarct volume in experimental focal stroke. Brain Res. 1992; 574(1-2):171–7.
110. Nikić I, Merkler D, Sorbara C, Brinkoetter M, Kreutzfeldt M, Bareyre FM, et al. A reversible form of axon damage in experimental autoimmune encephalomyelitis and multiple sclerosis. Nat Med. 2011;17(4):495–9. doi:10.1038/nm.2324. Epub 2011 Mar 27.
111. Bragin DE, Zhou B, Ramamoorthy P, Müller WS, Connor JA, Shi H. Differential changes of glutathione levels in astrocytes and neurons in ischemic brains by two-photon imaging. J Cereb Blood Flow Metab. 2010;30(4):734–8. doi:10.1038/jcbfm.2010.9. Epub 2010 Jan 27.
112. Bath PM, Macleod MR, Green AR. Emulating multicentre clinical stroke trials: a new paradigm for studying novel interventions in experimental models of stroke. Int J Stroke. 2009;4(6):471–9. doi:10.1111/j.1747-4949.2009.00386.x.
113. Macleod MR, Fisher M, O'Collins V, Sena ES, Dirnagl U, Bath PM, et al. Good laboratory practice: preventing introduction of bias at the bench. Stroke. 2009;40(3):e50–2. doi:10.1161/STROKEAHA.108.525386. Epub 2008 Aug 14.

Index

M. Rodriguez-Porcel et al. (eds.), *Studies on Atherosclerosis*, Oxidative Stress in Applied Basic Research and Clinical Practice,
DOI 10.1007/978-1-4899-7693-2